A DICTIONARY OF THERMODYNAMICS

Also published in this series:
A Dictionary of Spectroscopy
R.C. Denney

A Dictionary of Chromatography
R.C. Denney

In preparation:
A Dictionary of Electrochemistry
C.W. Davies and A.M. James
A Dictionary of Surface Chemistry
R.J. Breakspere and T.A. Egerton

A DICTIONARY OF THERMODYNAMICS

A.M. JAMES
M.A., D.Phil., D.Sc., C. Chem., F.R.I.C.
(*Professor of Physical Chemistry, Bedford College, University of London*)

M

First published 1976 by

THE MACMILLAN PRESS LTD

London and Basingstoke
Associated companies in New York, Dublin,
Melbourne, Johannesburg and Madras

SBN 333 18753 9

Photosetting by Thomson Press (India) Limited, New Delhi
and printed in Great Britain by Lowe & Brydone (Printers)
Ltd., Thetford, Norfolk

PREFACE

The idea of a dictionary of thermodynamics arose as a result of difficulties encountered during 25 years' experience of teaching the subject at all levels in university courses. At the start of a study of thermodynamics, students are not familiar with the language, while at the end of the course they require a thumb-nail sketch of the important concepts. In subsequent years, including periods of postgraduate study, students still need reminding of the basic concepts, equations, formulae, etc. and also their applications. While all these aspects are well and adequately covered in the many standard textbooks on thermodynamics, nevertheless in such books it is often impossible to locate readily a given topic and then to disentangle it from the context of the chapter. The fulfilment of these various needs has been uppermost in my mind during the writing of the book, in which the entries are listed in alphabetical sequence rather than by any of the methods arising from any conventional treatment of the subject.

Many thermodynamic courses and textbooks are devoted to the so-called 'rigorous' treatment of the subject, which generally means the derivation of all the standard equations. At the end, students have been treated to a course of mathematical gymnastics but they do not necessarily understand the significance or the application of the various concepts or equations to real systems. In this book, equations are quoted without derivation and their applications and limits are discussed; to all practising scientists this is a very important aspect.

The book is written essentially for scientists who have a minimum of G.C.E. 'A' level, or its equivalent, in a physical science subject. Biologists, Biochemists, Chemists, The Medical Profession,Geologists, Microbiologists, Pharmacists, Physiologists, Physicists and perhaps other scientists will, it is hoped, find this a useful reference book, which in addition to definitions of the various quantities, etc. also contains much useful tabulated thermodynamic data and provides access to standard reference texts for more detailed treatment. While not specifically aimed at schools, it is hoped that copies of the book will be available in the library for more advanced students.

Comments and suggestions for improvements would be most welcome.

I am indebted to Dr. W.H. Lee of Surrey University for many helpful discussions during the early stages of the preparation of the manuscript. The writing of any book consumes a large amount of time which would otherwise have been spent with the family. My sincere thanks are due to my wife, Dr. Mary P. Lord, for her understanding, patience, help and encouragement during the preparation of the manuscript.

1976

A.M.J.

USE OF THE DICTIONARY

Entries are arranged in alphabetical sequence with full cross-referencing where there is more than one acceptable or recognised name.

Symbols and abbreviations are used in the text without definition; reference should be made to the list of Principal Symbols on page viii.

In the text, a word in italics followed by (q.v.) indicates a reference to another entry which would be of help; in some instances the reference to another entry is in parentheses. Thus in the sentence "The *work* (q.v.) done during the isothermal reversible expansion of a perfect gas (see *reversible process*) is equal to the heat absorbed by the gas", the reader is referred to entries 'work' and 'reversible process' for further information. A word in italics, followed by†, e.g. *concentration cell*†, indicates a reference to that entry in a companion Dictionary of Electrochemistry by C.W. Davies and A.M. James.

More detailed treatments of some of the entries can be found in standard textbooks; where appropriate, relevant references are indicated at the end of an entry by a simple code for the name of the author. All the books with the coding are listed in the Bibliography.

PRINCIPAL SYMBOLS

A Helmholtz free energy; Debye–Hückel constant
B Debye–Hückel constant
C coulomb
C_p,C_V Heat capacity at constant pressure, volume
E Electromotive force of cell
E(O,R),E(X^+X) Electrode potential
F Faraday constant
G Gibbs free energy function
$\bar{G}_i$ Partial molar free energy of *i*th component
H Enthalpy or heat content
I Ionic strength; Moment of inertia
J joule
$\bar{J}$ Partial derivative
K kelvin
K Equilibrium constant with subscripts p,x, therm, s as appropriate
K_a,K_b Acid and base ionisation constants
K_w Ionisation constant of water
L_e,L_f Molar heat of vaporisation, fusion
M_r(X) Relative molecular mass of X
N Total number of particles or molecules in a system
N newton (SI unit of force)
N_A Avogadro constant
P Total pressure of system
P^* Fugacity
Q Partition function
R Gas constant
S Entropy
T Temperature/K
U Internal energy
V Volume
V volt
W Statistical probability
$\bar{X}_A$ Partial molar X of A

a_A,a_B,a_i Activity of A,B or *i*th component
a_+,a_-,$a_\pm$ Activity of cation, anion, mean ionic activity

c	Number of components (phase rule)
c_A	Concentration of A/mol dm^{-3}
e	Electron
f	Activity coefficient (mole fraction basis); Number of degrees of freedom (phase rule)
g	Gravitational constant; Statistical weight; Osmotic coefficient
g	gramme
k	Boltzmann constant
k	kilo (prefix), eg. kg = kilogramme = 10^3 g
k_c, k_e	Cryoscopic, ebullioscopic constant
l_e, l_f	Heat of vaporisation, fusion per kg
m	metre, milli (prefix), e.g. mm = millimetre = 10^{-3} m
m_A	Molality of A
$m_+, m_-, m_\pm$	Molality of cation, anion, mean ionic molality
n_A, n_B	Number of molecules of A,B in system
n_r	Number of particles in rth energy level
p	Number of phases (phase rule); Pressure above solution
p_i	Partial pressure of i in system
$p^\ominus$	Vapour pressure of pure solvent
p_i^*	Partial fugacity of ith component
$q, đq$	Heat absorbed by system
s	Second
t_+, t_-	Transport number of cation, anion
$w, đw$	Work done by system
w_A, w_B	Weight of A,B
x_A, x_B	Mole fraction of A,B in solution
y_A, y_B	Mole fraction of A,B in gaseous phase
z_A, z_B	Valence of A,B
α	Degree of association or dissociation
γ	Ratio of heat capacities C_p/C_V; Surface tension
$\gamma_i, \gamma_+, \gamma_-$	Activity coefficient (molality scale)
$\gamma_\pm$	Mean ionic activity coefficient
ε	Permittivity
ε_r	Energy of particle in rth energy level
μ	micro (prefix), e.g. μm = micrometre = 10^{-6} m
μ_A, μ_B	Chemical potential of A,B
μ_{JT}	Joule–Thomson coefficient

Principal symbols

ν, ν_+, ν_-	Number of ions, cations, anions formed from 1 mole of electrolyte, $\nu = \nu_+ + \nu_-$
π	Ratio of circumference to diameter of circle = 3.14159; Osmotic pressure
ρ	Density
σ	Symmetry number; Area
Γ_A	Surface excess concentration of A
Δ	Increase in thermodynamic function, e.g. $\Delta X = X_2 - X_1$
ΔX_f	Increase in X for formation of compound from its elements

Superscripts

$\ominus$	Indicating a standard value of a property

Subscripts

A,B...	Referring to substances A,B...
i	Referring to typical ionic species i, or component
p,V,T,S...	Indicating constant pressure, volume, temperature, entropy, etc.
f,e,s,t	Referring to fusion (or formation), evaporation, sublimation, transition
$+, -$	Referring to positive or negative ion
1,2	Referring to different systems or states of system
g,l,s (or c)	Referring to gaseous, liquid or solid states, respectively

Other abbreviations

b.p.	Boiling point
f.p.	Freezing point
m.p.	Melting point
v.p.	Vapour pressure
e.m.f.	Electromotive force
$\neq$	not equal to
$\equiv$	identically equal to
$\approx$	approximately equal to
$\propto$	proportional to
∞	infinity
$<$	smaller than
$>$	larger than
$\leqslant$	smaller than or equal to
$\geqslant$	larger than or equal to

$\ll$	much smaller than
$\gg$	much larger than
$p!$	factorial $p = 1 \times 2 \times 3 \ldots \times (p-1) \times p$
$\sum_i$	sum of i terms
$\prod_i$	product of i terms
$f(x)$	function of x
$\frac{\mathrm{d}f}{\mathrm{d}x}, \mathrm{d}f/\mathrm{d}x,$	differential coefficient of $f(x)$ with respect to x
$\left(\frac{\partial f}{\partial x}\right)_{y\ldots}$	partial differential coefficient of $f(x,y\ldots)$ with respect to x when $y\ldots$ are held constant
$\mathrm{e}^x, \exp(x)$	exponential of x
e	base of natural logarithms
$\ln x$	natural logarithm of x
$\log x$	common logarithm of x (to base 10), $\ln x = 2.303 \log x$

A

Absolute molar entropy

The absolute molar entropy of a substance, $S^{\ominus}(T)$, is evaluated from heat capacity data by the equation:

$$S^{\ominus}(T) = S_0^{\ominus} + \int_0^T C_p^{\ominus}\, \mathrm{d} \ln T + \sum \frac{\Delta H(\text{trans})}{T(\text{trans})}$$

For substances conforming to the *third law of thermodynamics* (q.v.), $S_0^{\ominus} = 0$.

Acids and bases

Arrhenius defined an acid as a compound which dissociated in solution to yield a hydrogen ion, and a base as a compound which yielded an hydroxyl ion. On this basis the process of neutralisation is simply

$$H^+ + OH^- \longrightarrow H_2O$$

This definition is, however, limited to the use of water as the solvent.

Table A.1. Conjugate acids and bases by Lowry–Brønsted definition

Acid (↑ Increasing strength of acid)		Proton +		Base (↓ Increasing strength of base)
$HClO_4$	$\rightleftharpoons$	H^+	+	ClO_4^-
HBr	$\rightleftharpoons$	H^+	+	Br^-
H_2SO_4	$\rightleftharpoons$	H^+	+	HSO_4^-
H_3O^+	$\rightleftharpoons$	H^+	+	H_2O
HSO_4^-	$\rightleftharpoons$	H^+	+	SO_4^{2-}
H_3PO_4	$\rightleftharpoons$	H^+	+	$H_2PO_4^-$
$[Co(NH_3)_4(H_2O)_2]^{3+}$	$\rightleftharpoons$	H^+	+	$[Co(NH_3)_4H_2O{\cdot}OH]^{2+}$
CH_3COOH	$\rightleftharpoons$	H^+	+	CH_3COO^-
NH_4^+	$\rightleftharpoons$	H^+	+	NH_3
HCO_3^-	$\rightleftharpoons$	H^+	+	CO_3^{2-}
H_2O	$\rightleftharpoons$	H^+	+	OH^-
NH_3	$\rightleftharpoons$	H^+	+	NH_2^-
CH_3OH	$\rightleftharpoons$	H^+	+	CH_3O^-
$(C_6H_5)_2NH$	$\rightleftharpoons$	H^+	+	$(C_6H_5)_2N^-$

Lowry and Brønsted (1923) independently put forward a general definition of acids and bases which has been of great importance in unifying the dissociation of acids and bases, the hydrolysis of salts and the catalytic action of acids and bases. In this, an acid is defined as a substance which can give up a proton, and a base as one that can accept a proton. An acid and a base are conjugate when related by the equation

$$\text{Acid} \rightleftharpoons \text{Proton} + \text{Base}$$

$$\text{A} \rightleftharpoons \text{H}^+ + \text{B}$$

An acid can only give up a proton if a base is present to accept it; hence, two acids and two bases must take part in any reaction involving transfer, i.e.

$$A_1 + B_2 \rightleftharpoons A_2 + B_1$$

Free protons do not exist in solution; thus a potential acid can only function as such when it is the solute only if the solvent is a proton acceptor. Conversely for a base. Thus the acidic or basic function of a species cannot become apparent unless the solvent is capable of showing basic or acidic properties, respectively, e.g.

$$\underset{A_1}{CH_3COOH} + \underset{\text{solvent, } B_2}{H_2O} \rightleftharpoons \underset{A_2}{H_3O^+} + \underset{B_1}{CH_3COO^-}$$

$$\underset{B_1}{CN^-} + \underset{\text{solvent, } A_2}{H_2O} \rightleftharpoons \underset{A_1}{HCN} + \underset{B_2}{OH^-}$$

in which water can act, according to circumstances, as an acid or as a base. Thus the original Arrhenius theory, which applied to aqueous solutions only is extended to cover more substances. The nature of the solvent is thus important in deciding whether a substance will show acidic or basic properties in solution. There are three main types of solvent: (1) protogenic solvents, proton donors with acidic properties, e.g. glacial acetic acid; (2) protophilic solvents, proton acceptors with basic properties, e.g. ammonia; (3) aprotic solvents, which can neither accept nor donate protons, e.g. toluene, solutes dissolved in such solvents cannot show either acidic or basic properties. Water, alcohols, etc., which can be either protogenic or protophilic according to circumstances, are said to be amphiprotic.

G.N. Lewis extended the definition still further in his definition of an acid as a substance which can accept a pair of electrons from a donor substance, the base. Neutralisation is thus the formation of a covalent bond in which both electrons of the shared pair are provided by the base, e.g.

$$\underset{\text{acid}}{A} + \underset{\text{base}}{:B} \rightleftharpoons \underset{\substack{\text{coordination}\\ \text{compound}}}{A:B}$$

$$\begin{array}{c} Cl \quad H \qquad\qquad Cl\ H \\ Cl:\ddot{B} + :\ddot{N}:H \rightleftharpoons Cl:\ddot{B}:\ddot{N}:H \\ Cl \quad H \qquad\qquad Cl\ H \end{array}$$

Donors which share electron pairs with cations are called ligands. The theory permits the classification as bases of such 'neutral' substances as H_2O, CO, Cl^-, SO_4^{2-}; the only qualification is that the species possesses an unshared electron pair. Lewis acids include metal ions, BF_3, $AlCl_3$, SiF_4 and any other electron pair acceptors. The relative strengths of Lewis acids vary widely with the strength of the base; thus, according to the Brønsted theory, OH^- is a stronger base than is ammonia, but, according to the Lewis theory, OH^- is a weaker base than is ammonia when reacting with Ag^+, although it is stronger than ammonia when reacting with H^+. There are some obvious examples of acids (e.g. HCl, H_2SO_4) which do not naturally fit the Lewis definition since they cannot plausibly accept electrons.

The coordination compound A:B has been formed from two reactants, each capable of an independent existence in solution, by the sharing of the electrons of the donor. Thus it is easy to distinguish this process from an ordinary oxidation–reduction process:

$$\text{Oxidised state} + n\,\text{e} \rightleftharpoons \text{Reduced state}$$

in which electrons are transferred between donor and acceptor and in which the resulting species have an independent existence.

The *equilibrium constant* (q.v.) for the general reaction, defined

$$K = \frac{[A:B]}{[A]\,[:B]}$$

where [ ] refers to the active masses, is called a stability or formation constant.

If two ligands can add to the acceptor (e.g. in the silver–ammonia complex), the second step is

$$A:B + :B \rightleftharpoons B:A:B$$

and the formation constant is

$$K = \frac{[\mathrm{B:A:B}]}{[\mathrm{A:B}][\mathrm{:B}]}$$

If a ligand can add to a series of acceptors (e.g. metal–ammonia complexes), the formation constant is largest for the reaction leading to the greatest amount of coordination compound; the greater the stability, the greater the formation constant.

The Lewis acids of principal interest in biological sciences are the metal ions, e.g. in coordination compounds such as haemoglobin, chlorophyll, the vitamin B_{12} group, cytochromes and metalloenzymes.

Dissociation constants

The dissociation of an acid HA in water is written

$$\mathrm{HA + H_2O \rightleftharpoons H_3O^+ + A^-}$$

for which

$$K_a = \frac{a(\mathrm{H_3O^+})\, a(\mathrm{A^-})}{a(\mathrm{HA})}$$

and of the base A^-,

$$\mathrm{A^- + H_2O \rightleftharpoons HA + OH^-}$$

for which

$$K_b = \frac{a(\mathrm{HA})\, a(\mathrm{OH^-})}{a(\mathrm{A^-})}$$

Hence, $K_a K_b = a(\mathrm{H_3O^+})\, a(\mathrm{OH^-}) = K_w = 10^{-14}\ \mathrm{mol^2\ dm^{-6}}$, i.e. the conjugate base (acid) of a strong acid (base) is weak. K_a may be written in terms of concentrations and *activity coefficients* (q.v.) of the various species:

$$K_a = \frac{c(\mathrm{H_3O^+})\, c(\mathrm{A^-})}{c(\mathrm{HA})} \times \frac{\gamma(\mathrm{H_3O^+})\, \gamma(\mathrm{A^-})}{\gamma(\mathrm{HA})} = K_c\, \frac{\gamma(\mathrm{H_3O^+})\, \gamma(\mathrm{A^-})}{\gamma(\mathrm{HA})}$$

where the equilibrium constant, K_c, can be obtained from the original concentration of acid and the equilibrium concentration of one product. In dilute solution $\gamma(\mathrm{HA}) = 1$ and $\gamma(\mathrm{H_3O^+})$ and $\gamma(\mathrm{A^-})$ can be obtained from the *Debye–Hückel activity equation* (q.v.), whence

$$\log K_a = \log K_c + 2\, A\, I^{1/2}$$

The graph of $\log K_c$ against $I^{1/2}$ is linear (figure A.1) and of intercept $\log K_a$.

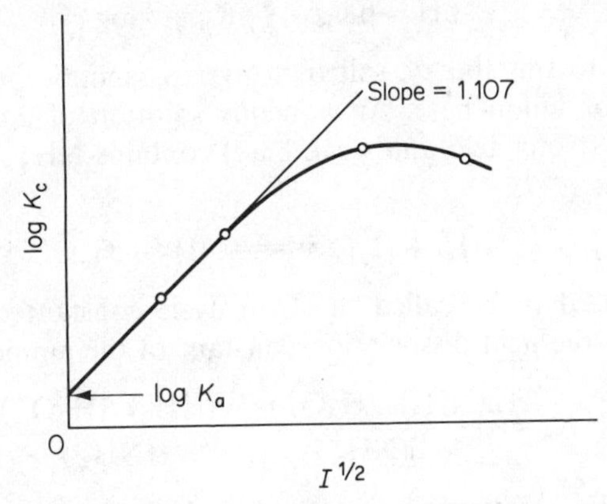

Figure A.1 Plot of log K_c against $I^{1/2}$ for a weak acid.

Alternatively, if α is the degree of dissociation and c/mol dm^{-3} is the initial concentration of acid,

$$K_a = \frac{\alpha^2 c}{1-\alpha} \times \frac{\gamma(H_3O^+)\,\gamma(A^-)}{\gamma(HA)}.$$

If activity coefficients are neglected, the classical dissociation constant is given by Ostwald's dilution law:

$$K_c = \alpha^2 c/(1-\alpha)$$

Dissociation constants provide a measure of the strengths of acids and bases; the larger K_a, the stronger the acid. Values of K_a are often so small that they are tabulated (Table A.X, p. 257) as pK_a values, where pK_a = $-\log K_a$; thus for acetic acid at 298 K, $K_a = 1.754 \times 10^{-5}$ mol dm^{-3} and pK_a = 4.756.

In the ionisation of polyprotic (polybasic) acids, each successive proton is removed from an ion of increased negative charge and so K_a for each step is less than that for the preceding step: e.g. for phosphoric acid $K_{a(1)} = 1.1 \times 10^{-2}$, $K_{a(2)} = 7.5 \times 10^{-8}$ and $K_{a(3)} = 4.8 \times 10^{-13}$ mol dm^{-3}.

For most organic acids, α is small and is approximately given by $\alpha = (K_a/c)^{1/2}$; hence, for a weak acid, $c(H^+) = c\alpha = (K_a\,c)^{1/2}$. Assuming $a(H^+) \approx c(H^+)$ (valid except in strongly acid solution), the pH of a solution of a weak acid is

$$\text{pH} = -\log c(H^+) = \text{p}K_a - \tfrac{1}{2}\log c$$

and that of a weak base is

$$pH = pK_w - \tfrac{1}{2}pK_b - \tfrac{1}{2}\log c$$

According to this theory, salt hydrolysis is simply the dissociation of a cation acid or anion base. An aqueous solution of ammonium chloride (the salt of a strong acid and weak base) contains NH_4^+, which reacts with the water thus:

$$NH_4^+ + H_2O \rightleftharpoons NH_3 + H_3O^+$$

Thus what used to be called the hydrolysis constant of the ammonium salt is clearly the acid dissociation constant of the ammonium ion:

$$K_h = K_a = \frac{a(NH_3)\,a(H_3O^+)}{a(NH_4^+)} \approx \frac{c(NH_3)\,c(H_3O^+)}{c(NH_4^+)} \approx \frac{c\alpha^2}{1-\alpha}$$

where c/mol dm^{-3} is the concentration of salt and α, the extent of hydrolysis, is given by

$$\alpha = \{K_a/c\}^{1/2} = \{K_w/K_b\,c\}^{1/2}$$

whence the pH of such a solution is

$$pH = -\log c\alpha = \tfrac{1}{2}\,pK_w - \tfrac{1}{2}\,pK_b - \tfrac{1}{2}\log c$$

Similarly, for the salt of a weak acid and strong base,

$$pH = \tfrac{1}{2}\,pK_w + \tfrac{1}{2}\,pK_a + \tfrac{1}{2}\log c$$

Thus the pH of a solution of the salt of a weak (strong) acid and a strong (weak) base is greater (less) than 7, i.e. alkaline (acidic) and increases (decreases) with increasing concentration of salt. For the salt of a weak acid and a weak base, the pH, given by

$$pH = \tfrac{1}{2}\,pK_w + \tfrac{1}{2}\,pK_a - \tfrac{1}{2}\,pK_b$$

depends on the relative strengths of the acid and base and is independent of concentration.

See also Dissociation constant; Potentiometric titration†; and Ad, B, G, G & S, W & W.

Acid–base titrations

See Potentiometric titration†.

Activity

Activity, a, is a function introduced by G.N. Lewis to aid the treatment of real systems. Like *fugacity* (q.v.), the activity permits the correlation of changes in *chemical potential* (q.v.) with experimentally measured quantities, such as concentration, through a relationship formally equivalent to that for an ideal system. Activity is defined as the ratio of the fugacity in a given state to the fugacity in a standard state, thus:

$$RT \ln a_i = \mu_i - \mu_i^\ominus = RT \ln \frac{p^*}{[p_i^*]^\ominus} \tag{A.1}$$

The use of the activity function for solids, liquids and gases is preferred to the fugacity, since it does not require comparison with an infinitely attenuated gas. For a pure gas, the standard state is that of unit fugacity; hence, $a_i = p_i^*$. Activity, defined as the ratio of fugacities, is dimensionless and, as such, its numerical value is meaningless unless a standard reference state is defined. Comparison of equation (A.1) with that for an ideal system:

$$RT \ln x_i = \mu_i - \mu_i^\ominus$$

suggests the introduction of an *activity coefficient* (q.v.), f_i, defined as $f_i = a_i/x_i$. The deviation of f_i from unity is a measure of the deviation of the system from ideal behaviour. When the composition variable is molality, the activity coefficient, $\gamma_i = a_i/m_i$, is the practical activity coefficient, since molality is the most common method of expressing concentrations, particularly of dilute solutions.

Activity is a function of the temperature, pressure and composition of the system:

$$\left(\frac{\partial \ln a_i}{\partial P}\right)_{T,x} = \frac{\bar{V}_i}{RT}$$

$$\left(\frac{\partial \ln a_i}{\partial T}\right)_{P,x} = \frac{\bar{H}_i - \bar{H}_i^\ominus}{RT^2}$$

where $\bar{H}_i$ and $\bar{H}_i^\ominus$ are the partial molal enthalpy values in the given and standard states, respectively.

Since no method has been devised to determine individual ion activities (a_+, a_-), the mean ionic activity $a_\pm$ for dissociated electrolytes is defined:

$$a_\pm = a_2^{1/\nu} = (a_+^{\nu+} \, a_-^{\nu-})^{1/\nu}$$

where a_2 is the activity of the undissociated electrolyte.

The variation of a or $a_\pm$ for sucrose and various electrolytes is shown in figure A.2.

Thermodynamic *equilibrium constants* (q.v.) are defined in terms of the

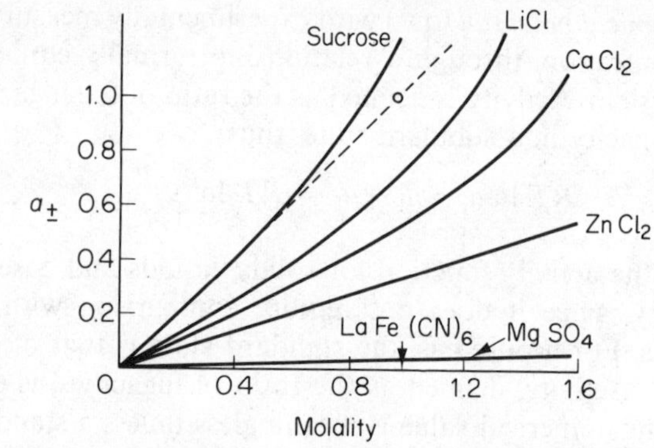

Figure A.2 Plot of mean ionic activity of various salts in water against the molality at 25 °C. (Circle on dotted line represents standard state.)

activities of the products and reactants:

$$K_{\text{therm,}} = \frac{\Pi\, a\,(\text{products})}{\Pi\, a\,(\text{reactants})} = K_c \times \frac{\Pi\, \gamma\,(\text{products})}{\Pi\, \gamma\,(\text{reactants})}$$

Determination of activities
Activities can be obtained from the results of any of the methods available for the measurement of free energy changes.

(1) Vapour pressure of solute. In principle, measurements of the v.p. of the solute, p_B, is the simplest method of determining the activity of an electrolyte which is volatile enough to permit accurate measurements at low concentration. The graph of p_B against $m_\pm$ allows the calculation of the limiting slope ($= p_B^\ominus$) and, hence, $a_\pm$. This is not a very practical method.

(2) Solubility of a sparingly soluble salt. From solubility measurements of sparingly soluble salts in electrolyte solutions of known ionic strength, the *solubility product* (q.v.) and, hence, $\gamma_\pm$ and $a_\pm$ can be obtained. This method is limited to low concentrations of a few salts.

(3) EMF studies of *galvanic cells*† and *concentration cells*† for the measurement of standard *electrode potentials*†. The experimental data permits the calculation of the mean ionic activity and activity coefficients. The method is limited to strong electrolytes which can be formed in a reversible cell.

(4) *Vapour pressure* (q.v.) of the solvent. Measurements of the v.p. of the solvent afford the most accurate method of determining activities in concentrated solutions (> 0.1 mol kg^{-1}). The activity of the solvent is obtained from the observed partial v.p. and that of the pure solvent. The activity of

the solute in solution can be obtained from the activity of the solvent using the *Gibbs–Duhem equation* (q.v.).

(5) Depression of the *freezing point* (q.v.) of a solution. This method is applicable when (1) and (4) cannot be used on account of small differences of pressure between solvent and solution. The activity of the solvent, a_A, can be determined at its f.p. by the equation

$$\ln a_A = \frac{-L_f}{RT_0^2} \Delta T_f$$

and that of the solute, on application of the *Gibbs–Duhem equation* (q.v.),

$$d \ln a_B = \frac{L_f}{RT_0^2} \frac{d(\Delta T_f)}{m_B} = \frac{d(\Delta T_f)}{m_B k_f}$$

From a study of the variation of ΔT_f with m_B, a_B can be evaluated by graphical integration. These equations give a_A and a_B (or $a_\pm$ for electrolytes) at the f.p. of the solution; this is at a different temperature for each concentration but, for dilute solutions, the error is negligible.

(6) Elevation of the *boiling point* (q.v.) of a solution. Measurements of the elevation of the b.p. with concentration can be used to determine a_A and a_B; the equations are similar to those in (5).

See also Osmotic coefficient; and D & J, G, G & S, I, L & R.

Activity coefficient

Activity coefficient is the ratio of the *activity* (q.v.) to the concentration (or pressure) of a component i in the given state, i.e. $f_i = a_i/x_i$, $\gamma_i = a_i/m_i$ or, for gases, $f = P^*/P$. The activity coefficient, which approaches unity as the concentration (or pressure) approaches zero, is a measure of the departure of the system from ideal behaviour.

For electrolytes, γ_+ and γ_- cannot be measured separately; hence, the mean ionic activity coefficient $\gamma_\pm$ is defined

$$\gamma_\pm = [\gamma_+^{\nu+} \gamma_-^{\nu-}]^{1/\nu} = a_\pm/m_\pm$$

$\gamma_\pm$ decreases rapidly with increasing molality and, at higher concentrations, usually passes through a minimum value and then increases (figure A.3). The steepness of the initial decrease varies with the valence type of the electrolyte; the greater the product of the valence of the ions, the greater the deviation from ideal behaviour at a given concentration. For a given valence type, $\gamma_\pm$ is independent of the constituent ions at molalities less than 0.01 mol kg^{-1}; above this concentration, the curves separate and specific ion effects become apparent.

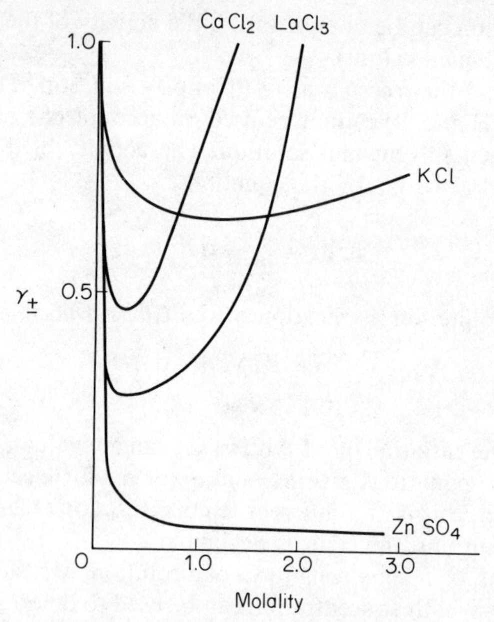

Figure A.3 Activity coefficients of electrolytes of different valence type.

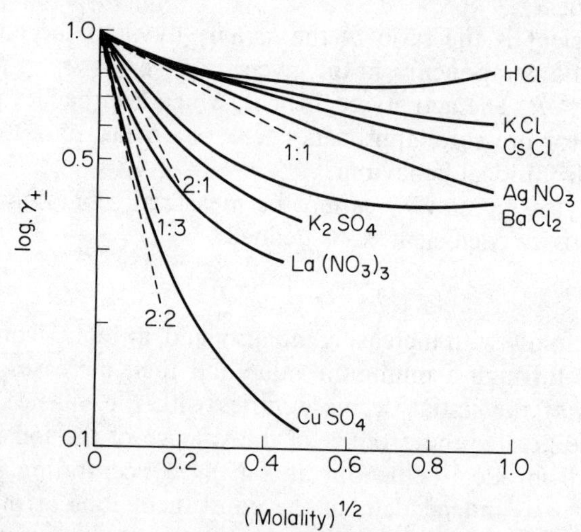

Figure A.4 Mean ionic activity coefficients as a function of molality. (Dotted lines show behaviour expected from Debye–Hückel limiting law.)

In dilute solutions (< 0.01 mol kg^{-1}), γ_i and $\gamma_\pm$ are given by the *Debye–Hückel activity equation* (q.v.) (figure A.4, broken lines):

$$\log \gamma_i = -A z_i^2 I^{1/2}$$

and

$$\log \gamma_\pm = -A z_+ z_- I^{1/2}$$

At higher ionic strengths, the complete equation is necessary.

The activity coefficient is related to the *osmotic coefficient* (q.v.), g, by the equation

$$1 - g = -\tfrac{1}{3} \ln \gamma$$

Activity coefficients can be determined by any of the methods available for activity determinations; results from different experimental methods are in good agreement.

See also Chemical potential; table A.VIII (p. 256); and G, I, K, R & S.

Activity solubility product
See Solubility product.

Adiabatic process
An adiabatic process or isentropic process is one in which no heat enters or leaves the system: $q = 0$, $\Delta S = 0$. If work is done by the system during such a process, there is a decrease in the internal energy and, hence, the system cools.

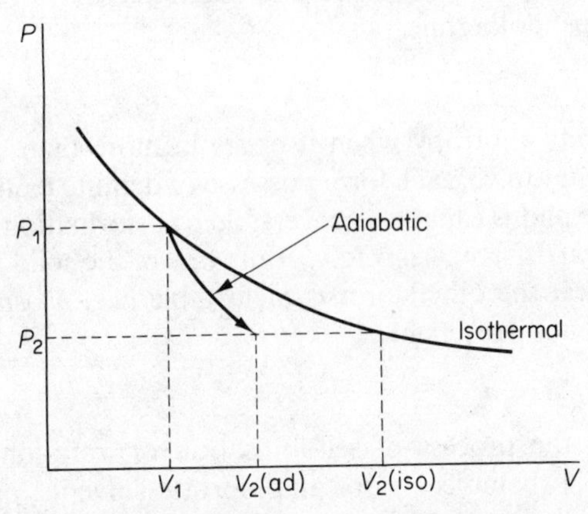

Figure A.5 Comparison of isothermal and adiabatic expansions of a gas.

Adiabatic process

For a reversible adiabatic process (e.g. expansion; see figure A.5) involving an ideal gas:

$$P_1 V_1^\gamma = P_2 V_2^\gamma = K \qquad (\gamma = C_p/C_V)$$

$$dU = -đw = -C_V \, dT = -P \, dV$$

$$C_p \ln(T_2/T_1) = R \ln(P_2/P_1) \quad \text{or} \quad T_2/T_1 = (P_2/P_1)^{R/Cp}$$

In an isothermal expansion (see *isothermal process*) of a perfect gas, *work* (q.v.) done by the system is paid for by heat absorbed. In an adiabatic process, this is not possible and work done is compensated for by a decrease in the internal energy of the gas and, hence, in its temperature. The volume then does not increase as much for a given pressure drop.

See also K.

Affinity

Affinity is the tendency of a reaction to proceed. The only satisfactory measure of the affinity of a reaction is the decrease in the Gibbs *free energy* (q.v.) function ($-\Delta G$) which accompanies the reaction. The reaction is possible if there is a decrease in the free energy, i.e. $\Delta G < 0$. Since

$$\Delta G = \Delta H - T\Delta S$$

the affinity of a reaction can be found from a knowledge of ΔH and ΔS.

Thomsen and Berthelot, from their observations that *exothermic processes* (q.v.) are more common and more vigorous than *endothermic processes* (q.v.), proposed that the heat produced in a chemical reaction should be regarded as its affinity. This is not a satisfactory definition of affinity, since endothermic reactions do occur; further, in reversible reactions, that occurring in one direction must be endothermic.

Allotropy

An element exhibits allotropy when it exists in more than one crystalline from. In *enantiotropy* (q.v.), each form possesses a definite temperature range of stable existence and is capable of reversible transformation into the other (e.g. *sulphur system*). In *monotropy* (q.v.), only one of the solid forms is stable at all temperatures; the other form is metastable (e.g. *phosphorus system*). The transformation is irreversible.

Autoprotolysis

Autoprotolysis is the process of self-ionisation or auto-ionisation which proceeds between two molecules of amphiprotic solvents, one as an acid and the other as a base, as in water:

$$H_2O + H_2O \rightleftharpoons H_3O^+ + OH^-$$

The extent of autoprotolysis is indicated by the *equilibrium constant* (q.v.), which, for water at 298 K, is

$$K_w = \frac{a(H_3O^+)\,a(OH^-)}{a^2(H_2O)}$$

Since pure liquid is the standard state of the solvent, its activity is unity, and the *ionic product of water* (q.v.) becomes

$$K_w = a(H_3O^+)\,a(OH^-) = 1.008 \times 10^{-14}\ \text{mol}^2\ \text{dm}^{-6}$$

The autoprotolysis constant for a given solvent depends on its acid strength, its basic strength and the relative permittivity. Autoprotolysis is exhibited only by solvents with both acidic and basic properties. It does not occur in aprotic solvents (e.g. benzene), which neither give up nor accept a proton.

Azeotrope

An azeotrope or azeotropic mixture is a mixture of two completely miscible liquids which cannot be separated by fractional distillation. The mixture distils over unchanged in composition at a fixed b.p., which may be greater or less than the b.p. of the separate components. It is not a simple chemical compound, as its composition and its b.p. vary with the external pressure. The separation of the two components may be effected by fractional crystallisation, by the addition of a third component to alter the characteristics of the system (e.g. benzene to ethanol/water), followed by distillation, or by the selective removal of one component (e.g. removal of water by silica gel).

See also Binary liquid mixture; Completely miscible liquid systems.

B

Babo's law

Babo's law states that the addition of a non-volatile solute to a volatile solvent in which it is soluble lowers the v.p. of the solvent in proportion to the amount of substance dissolved.

See also Boiling point; Freezing point; Raoult's law.

Bases

See Acids and bases

Beckmann thermometer

The Beckmann thermometer is used for measuring differences in temperature which are too small to be measured with an ordinary thermometer. This thermometer, with a range of 6 K and graduated in 0.01 K, registers temperature differences and not absolute values. The thermometer can be adjusted, by transferring mercury to or from a reservoir at the top of the capillary tube, to read at any desired temperature.

The Beckmann thermometer is used to determine the elevation of the *boiling point* (q.v.) or depression of the *freezing point* (q.v.) of a solvent on the addition of a non-volatile solute.

This thermometer is being replaced by *thermistors* (q.v.), which have the advantages of small heat capacity and greater sensitivity.

Binary liquid mixture

A binary liquid mixture is a two-component system in which both components are volatile liquids. On account of deviations from *Raoult's law* (q.v.) and *Henry's law* (q.v.), these liquid mixtures can be classified as *completely miscible liquid systems* (q.v.), *partially miscible liquids* (q.v.) and *immiscible liquids* (q.v.).

Binodal curve

See Three-component system.

Blagden's law

Blagden's law states that the lowering of the *freezing point* (q.v.) of a volatile solvent on the addition of a non-volatile solute is proportional to the concentration of the solute.

Boiling point

The b.p. of a liquid is the temperature at which the *vapour pressure* (q.v.) of the liquid is equal to the external pressure. The normal b.p. is the temperature at which the v.p. = 101 325 N m^{-2} (760 mmHg). A high (low) b.p. indicates large (small) intermolecular forces of attraction.

The direct consequence of the lowering of the v.p. of a volatile solvent A by a non-volatile solute B is that the b.p. of the solution is higher than that of the solvent.

The elevation of b.p. (ΔT_e), assuming ideal behaviour, is related to the v.p. above the solvent and the solution by the equation

$$\Delta T_e = \frac{RT_0^2}{L_e} \ln \frac{p^\ominus}{p}$$

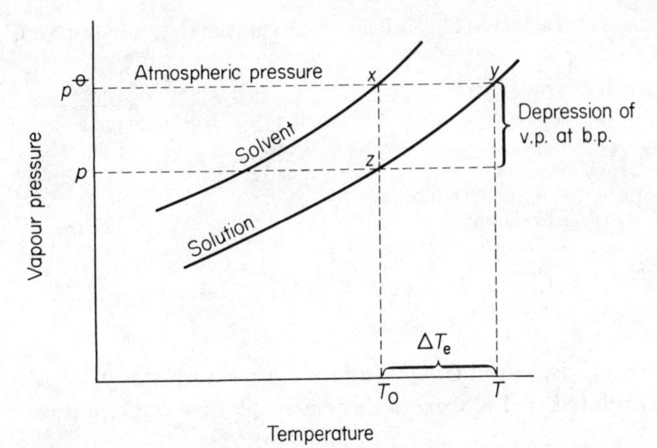

Figure B.1 The vapour pressure–temperature curves for solvent and solution near the b.p. of the solvent.

which is obtained by integrating the *Clausius–Clapeyron equation* (q.v.) between $y(p^{\ominus}, T)$ and $z(p, T_0)$ (figure B.1). Assuming the validity of *Raoult's law* (q.v.), ΔT_e is related to the concentration of solute by

$$\Delta T_e = \frac{RT_0^2}{L_e} x_B$$

It is more usual to use the molality of B, represented by m_B, rather than x_B, if n_A is the number of moles of solvent in 1 kg solvent, then

$$\Delta T_e = \left(\frac{RT_0^2}{L_e\, n_A}\right) m_B$$

The expression in parentheses is the molal ebullioscopic constant, k_e, for the given solvent and, with this notation,

$$\Delta T_e = k_e\, m_B$$

The b.p. elevation is clearly a function of the properties of the solvent and independent of any feature of the solute except its concentration; it is thus a *colligative property* (q.v.).

If deviations from ideal behaviour are taken into consideration, the *activity* (q.v.) of the solvent at the b.p. can be obtained from the equation

$$\ln a_A = \frac{-L_e}{RT_0} \Delta T_e$$

Table B.1. Molal ebullioscopic constants for various solvents

Solvent	b.p./°C	k_e/K kg mol^{-1}
Acetic acid	118.3	3.07
Acetone	56.1	1.71
Benzene	80.2	2.63
Carbon tetrachloride	76.8	5.0
Chloroform	61.5	3.77
Ethanol	78.3	1.2
Water	100.0	0.513

and hence a_B, using the *Gibbs–Duhem equation* (q.v.).

ΔT_e is related to the *osmotic pressure* (q.v.) by the equation

$$\Pi V = L_e \frac{\Delta T_e}{T_0}$$

From the experimental values of ΔT_e for solutions of known weight concentration, $M_r(B)$ can be calculated. All the equations are valid only for dilute solutions and, hence, the values of ΔT_e are small. These small temperature differences can be measured with a *Beckmann thermometer* (q.v.) or with *thermocouples* (q.v.) or *thermistors* (q.v.). In the experimental measurement of the b.p. of a solution, the measuring device must be immersed in the solution; it is therefore important to eliminate superheating. This has been achieved (compare the original Beckmann method): (a) by passing solvent vapour into the solution, thereby raising the solution to its b.p. (Landsberger's method), and (b) by boiling the solution and making it pump itself over the measuring device held in the vapour (Cottrell's method). The modern ebulliometer makes use of this latter principle, using a thermistor as the measuring device.

The experimental methods are limited to compounds of low relative molecular mass; otherwise ΔT_e is too small to be measured accurately. Anomalous values of $M_r(B)$ are obtained for solutes which undergo association or dissociation in the solvent.

See also Freezing point; Osmotic coefficient; and Bar, G & S, J & P.

Boiling point elevation

See Boiling point.

Boltzmann equation

The Boltzmann equation relates the *entropy* (q.v.) of a system to its degree of disorder or thermodynamic probability W, i.e. $S = f(W)$. For subsystems A and B and the system AB, $S_A = f(W_A)$, $S_B = f(W_B)$ and $S_{AB} = f(W_{AB})$; but entropy is an additive property, i.e.

$$S_{AB} = S_A + S_B$$

and, hence, $f(W_{AB}) = f(W_A) + f(W_B)$

Since any microstate of subsystem A can combine with any microstate of B to form a distinct microstate of system AB, it follows that $W_{AB} = W_A \times W_B$; the implication is that the function relating S and W is logarithmic, i.e.

$$S = k \ln W + S_0$$

According to the *Nernst heat theorem* (q.v.), $S_0 = 0$ and, hence, $S = k \ln W$. This is the Boltzmann equation; its use involves an analysis of the distribution of all the atoms and molecules among the various energy levels available to the system. This equation, which has a major role in many developments in *statistical thermodynamics* (q.v.), makes possible the calculation of all the thermodynamic properties of a system in terms of the mechanical properties (velocities, momenta, kinetic and potential energies) of the microstates.

See also Partition function; Spectroscopic entropy; and Kn, N, Ru.

Bomb calorimeter

The bomb calorimeter is used for the measurement of the *enthalpy* (q.v.) of combustion. A weighed amount of sample is placed in a dish in the heavy-walled bomb of corrosion-resistant alloy (figure B.2). The bomb is filled with O_2 at 30 atm pressure and immersed in a calorimeter containing a known weight of water, a stirrer, a thermometer and an electric heating coil. An electric current is passed through the wire spiral in the sample to initiate the reaction; the rise in temperature of the water due to the combustion is measured by a platinum resistance thermometer sensitive to 10^{-4} K. In a subsequent experiment the heat change is determined from the electrical energy required (supplied via the coil) to duplicate the temperature rise.

The experimental arrangement insulates the calorimeter from its surroundings (i.e. adiabatic calorimeter); the apparatus is calibrated with benzoic acid, $\Delta H_c^{\ominus} = -3226.41 \pm 0.33$ kJ mol^{-1}. Results with errors less than 0.01% are readily attainable.

This method, at constant volume, measures ΔU, from which the enthalpy change can be calculated:

Bomb calorimeter

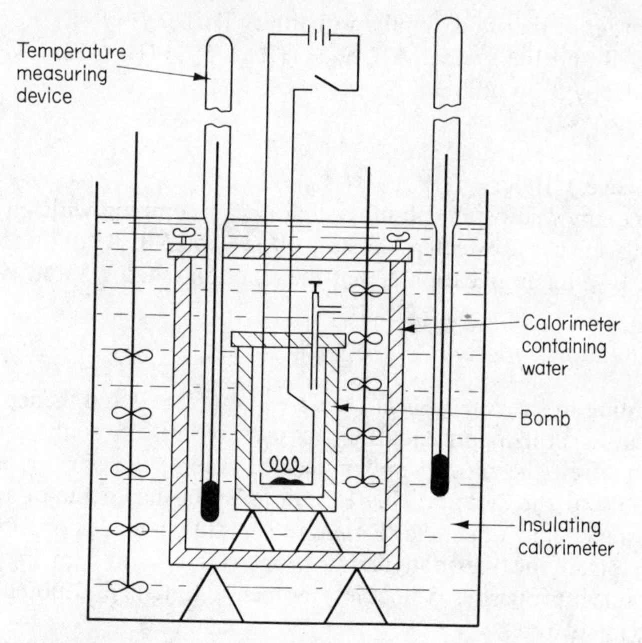

Figure B.2 Diagrammatic illustration of bomb calorimeter.

$$\Delta H = \Delta U + \Delta nRT$$

The bomb calorimeter can only be used for instantaneous reactions; otherwise it is impossible to correct for heat losses.

Bond energy

For a simple diatomic molecule, the bond energy is the heat energy required to break the molecule into the individual atoms. This process is known as dissociation and, in this instance, the bond-dissociation energy and the bond energy are identical. For polyatomic molecules, the bond energy is an intermediate value of the dissociation energies of a given bond in a series of different dissociating species.

Thus, for the O—H bond, the heat of dissociation depends on the nature of the species from which the H atom is being separated:

$$H_2O(g) \longrightarrow OH(g) + H(g) \quad \Delta H(298\ \mathrm{K}) = 501.87\ \mathrm{kJ}$$

$$OH(g) \longrightarrow O(g) + H(g) \quad \Delta H(298\ \mathrm{K}) = 423.38\ \mathrm{kJ}$$

These values are the bond-dissociation energies. The bond energy of the O—H bond is the average of these: ε(O—H) = 462.63 kJ.

The reaction

$$H_2O(g) \longrightarrow O(g) + 2H(g) \quad \Delta H(298\ K) = 925.25\ kJ$$

represents the breaking of two O—H bonds.

The bond energy refers to an intermediate average value of the dissociation energy of a given type of linkage for the dissociation of the gaseous compound into gaseous atoms.

Bond energies are of particular value in problems involving organic compounds; the bond energies are calculated from known heats of combustion and atomisation. ε(C—H) is, by definition, one-quarter of the heat of dissociation of gaseous methane into gaseous C and H; it is obtained from the following series of reactions:

$$CH_4(g) + 2O_2(g) \longrightarrow CO_2(g) + 2H_2O(l) \quad \Delta H = -890.4\ kJ$$

$$CO_2(g) \longrightarrow C(graphite) + O_2(g) \quad \Delta H = 393.5\ kJ$$

$$2H_2O(l) \longrightarrow 2H_2(g) + O_2(g) \quad \Delta H = 571.6\ kJ$$

$$2H_2(g) \longrightarrow 4H(g) \quad \Delta H = 871.8\ kJ$$

$$C(graphite) \longrightarrow C(g) \quad \Delta H = 718.4\ kJ$$

whence, using *Hess's law of constant heat summation* (q.v.),

$$CH_4(g) \longrightarrow C(g) + 4H(g) \quad \Delta H = 1654.9\ kJ$$

and ε(C—H) = 1654.9/4 = 423.6 kJ.

Values of bond energies for bonds involving C vary considerably according to the value taken for the heat of sublimation of C, which cannot be measured directly at room temperature. Using this value of ε(C—H) and the known heats of combustion of other organic molecules, a similar procedure permits the calculation of average bond energies for the C—C, C═C, C≡C, C═O, C—X bonds; thus, for

$$C_xH_y(g) \longrightarrow xC(g) + yH(g) \quad \Delta H = Y\ kJ$$

$$\varepsilon(C{-}C) = (Y - y\varepsilon(C{-}H))/x.$$

Although the concept of bond energy is very useful, it is no substitute for accurate calorimetric data when available. Approximate enthalpies of formation, atomisation and, hence, resonance energies may be calculated from tables of bond energies (table A.II, p. 252).

See also Da.

Born–Haber cycle

The Born–Haber cycle is the application of *Hess's law of constant heat summation* (q.v.) to the calculation of crystal lattice energies, which cannot be obtained by a direct experimental method. The lattice energy is the energy absorbed in a process such as

$$\mathrm{NaCl(c)} \longrightarrow \mathrm{Na^+(g)} + \mathrm{Cl^-(g)}$$

the energy for this particular reaction can be obtained by the indirect route:

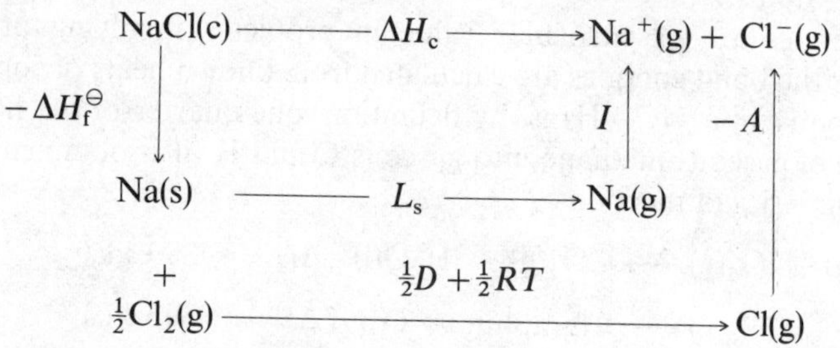

$$\begin{array}{lccc}
\mathrm{NaCl(c)} & \xrightarrow{\qquad \Delta H_c \qquad} & \mathrm{Na^+(g)} + & \mathrm{Cl^-(g)} \\
\downarrow -\Delta H_f^\ominus & & \uparrow I & \uparrow -A \\
\mathrm{Na(s)} & \xrightarrow{\qquad L_s \qquad} & \mathrm{Na(g)} & \\
+ & & & \\
\tfrac{1}{2}\mathrm{Cl_2(g)} & \xrightarrow{\qquad \frac{1}{2}D + \frac{1}{2}RT \qquad} & & \mathrm{Cl(g)}
\end{array}$$

where $\Delta H_f^\ominus$ is the standard enthalpy of formation of NaCl (−411 kJ), L_s is the heat of sublimation of Na (108.8 kJ), D the dissociation energy of Cl_2 ($\frac{1}{2}D + \frac{1}{2}RT = 122.2$ kJ), I the ionisation potential of Na (495.8 kJ) and A the electron affinity of Cl (365.3 kJ). The lattice energy is thus

$$\Delta H_c = -\Delta H_f^\ominus + L_s + \tfrac{1}{2}D + \tfrac{1}{2}RT + I - A$$

for NaCl, $\Delta H_c = 772$ kJ mol^{-1}.

Dissociation energies, D, are generally energies rather than enthalpies, and the term $\frac{1}{2}RT$ is necessary to correct this. For systems involving Br_2 and I_2, $\Delta H_f^\ominus$ is given in terms of the standard states, which are liquid and solid, respectively; it is thus necessary to add half the heat of vaporisation of bromine or half the heat of sublimation of iodine.

See also Da.

Buffer solution

A buffer solution is a solution which maintains a nearly constant *pH* (q.v.) despite the addition of small amounts of acid or alkali. Thus a solution of NaCl in water (pH ≈ 7) cannot maintain its pH as H^+ or OH^- is added. Many solutions, however, have a considerable reserve to remove added H^+ or OH^- and, hence, maintain a constant pH; such systems are said to show buffer action.

Most buffer solutions consist of one or more weak acids and their conjugate bases (see *acids and bases*); buffer solutions containing a weak base and conjugate acid, e.g. NH_4OH and NH_4Cl, are less common. For the solution of a weak acid and its conjugate base in water:

$$NaA \longrightarrow Na^+ \quad + A^-$$

$$H_2O + HA \rightleftharpoons H_3O^+ + A^-$$

any added H^+ is removed by the reverse reaction to give undissociated acid, while added OH^- is removed by the free acid:

$$HA + OH^- \longrightarrow A^- + H_2O$$

Continual addition of acid or base eventually results in complete displacement of equilibrium one way, and buffer action ceases. From the definition of *dissociation constant* (q.v.), it follows that

$$\mathrm{pH} = \mathrm{p}K_a + \log \frac{c(A^-)}{c(HA)} + \log \frac{\gamma(A^-)}{\gamma(HA)}$$

$$= \mathrm{p}K_a + \log \frac{c(A^-)}{c(HA)} - A\,I^{1/2} + C\,I$$

where $c(A^-)$ and $c(HA)$ are the total concentrations of salt and free acid, respectively, and A and C are the Debye–Hückel constants. The simplified version of this equation, without the activity correction terms, generally known as the Henderson equation, shows that the pH of a buffer solution is determined by (a) the $\mathrm{p}K_a$ of the weak acid and (b) the ratio $c(A^-)/c(HA)$; it gives no information, however, about the buffer capacity or the concentrations of acid and salt to use.

Buffer capacity, β, is measured by dB/dpH, where dB is the amount of strong base which when added to a buffer solution produces an increment dpH in the pH (it is the reciprocal of the slope of the pH–titration curve at a given point). In the effective buffer region (neglecting activity coefficients),

$$\beta = \frac{2.3\,a\,K_a\,c(H^+)}{[K_a + c(H^+)]^2}$$

i.e. β is directly proportional to a, the total concentration of free acid and salt. β is a maximum when $K_a = c(H^+)$, i.e. when $\mathrm{pH} = \mathrm{p}K_a$ and $c(A^-)/c(HA) = 1$; this is the mid-point of the neutralisation curve (figure B.3). β_{max} $(= 2.3\,a/4)$ is independent of K_a.

If $c(H^+)/c(HA)$ is increased or decreased by a factor of 10, i.e. $\mathrm{pH} = \mathrm{p}K_a \pm 1$, then, at these extreme values, $\beta = 2.3\,a/12$, which, although appreciable,

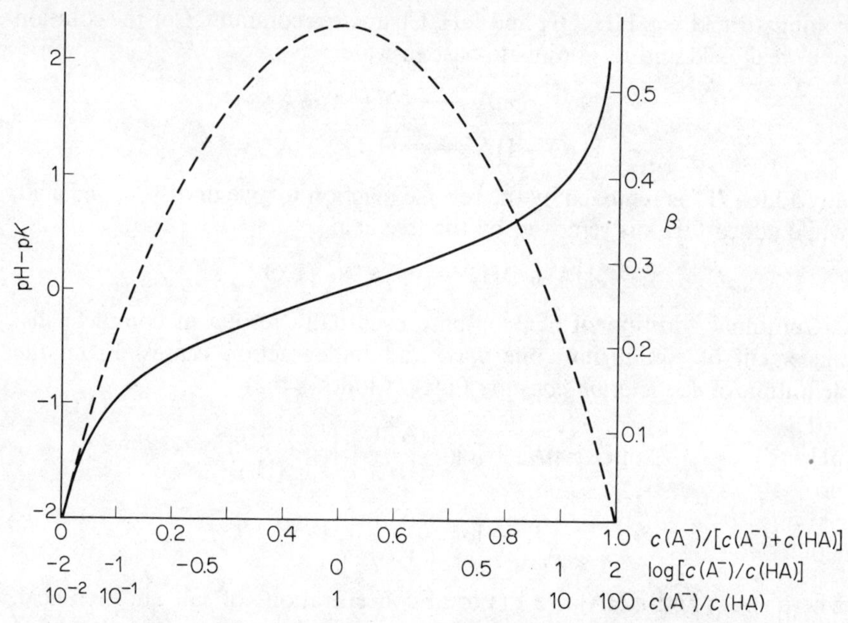

Figure B.3 Titration curve and buffer capacity.

is about $\frac{1}{3}\beta_{max}$; beyond these extremes, β decreases so rapidly that there is no buffer action.

For a given weak acid, the buffer solution is only of use for 1 pH unit on either side of the pK_a; hence, it is necessary to select an acid of pK_a nearest to the required pH. It is not necessary for the acid or salt to be a neutral molecule: e.g., in Sørensen's phosphate buffer solution,

$$\underset{(NaH_2PO_4\,:\,\text{acid})}{H_2PO_4^-} + H_2O \rightleftharpoons \underset{(Na_2HPO_4\,:\,\text{salt})}{HPO_4^{2-}} + H_3O^+$$

Then $c(A^-)/c(HA)$ is calculated from the Henderson equation; and the buffer solution of this ratio is then prepared. The effect of variation of ionic strength on the pH can be calculated from

$$pH = pK_a + \log[c(A^-)/c(HA)] - (2n-1)A\,I^{1/2} + C\,I$$

which applies to the nth stage of neutralisation of a polyprotic acid; hence,

$$d\,pH/dI^{1/2} = -(2n-1)\,A + 2C\,I^{1/2}$$

Thus the pH may increase or decrease with I, depending on conditions.

Table B.2. Buffer solutions

Solutions	pH range
HCl and KCl	1.0–2.2
Glycine and HCl	1.0–3.7
KHphthalate and HCl	2.2–3.8
Acetic acid and NaOH	3.7–5.6
KH_2PO_4 and NaOH	5.8–8.0
Boric acid and borax	6.8–9.2
Diethylbarbituric acid and sodium salt	7.0–9.2
Boric acid and NaOH	7.8–10.0
Na_2HPO_4 and NaOH	11.0–12.0

The data for many buffer solutions have been tabulated (table B.2); the pH values only apply to the concentrations quoted. Mixed buffer solutions, providing good buffer capacity over a wide range of pH values, can be obtained by using mixtures of weak acids, e.g. McIlvaine's buffer (pH 2.2 to 8.0), consisting of Na_2HPO_4 and citric acid, has five conjugate acid–base pairs.

Although solutions of strong acids and bases are not normally classified as buffer solutions, they have large buffer capacity at high concentrations; for a strong acid or base,

$$\beta = 2.3[c(H^+) + c(OH^-)]$$

Thus at pH 1 or 13, $\beta = 0.23$, in agreement with the relatively flat pH–neutralisation curves in the early stages.

See also Br, G, V.

C

Calorie

The calorie is the amount of heat required to raise the temperature of 1 gramme of water 1 K at 288 K. It is not coherent with the SI units and its use is discouraged. (1 calorie = 4.184 J.)

Carnot cycle

A Carnot cycle for a perfect gas is a typical reversible cycle of operations for the conversion of *heat* (q.v.) energy into *work* (q.v.). In such a cycle, performed in a reversible *heat engine* (q.v.), the maximum amount of work obtainable from a quantity of heat q_2 absorbed at temperature T_2 is given by

$$w_{\max} = \frac{q_2(T_2 - T_1)}{T_2}$$

where T_1 is the temperature of the heat reservoir which receives the heat which has not been transformed into work. When the temperature difference between the two isothermal stages of a Carnot cycle is a small amount, $\mathrm{d}T$, this may be written

$$\text{đ}w = q\,(\mathrm{d}T/T)$$

where q is the heat absorbed at the higher temperature T and $\text{đ}w$ is the total work done during the cycle.

Essentially the cycle consists of a sequence of 4 operations on n mole of the gas, for each of which the heat absorbed and work done by the gas are listed in table C.1 (see figure C.1).

$$\textit{Efficiency}\text{ (q.v.)} = \frac{w_{\max}}{q_2} = \frac{q_2 - q_1}{q_2} = \frac{nRT_2 \ln(p_1/p_2) - nRT_1 \ln(p_4/p_3)}{nRT_2 \ln(p_1/p_2)}$$

$$= \frac{T_2 - T_1}{T_2}$$

(since $p_1/p_2 = p_4/p_3$ from the ideal gas laws for isothermal and adiabatic changes). Hence,

$$q_2/T_2 - q_1/T_1, \quad \text{i.e.} \quad \Delta S_2 - \Delta S_1 = 0$$

Table C.1

Nature of process	Heat absorbed by gas	Work done by gas
1. Isothermal reversible expansion A to B	q_2	$w_1 = q_2 = nRT_2 \ln(p_1/p_2)$ = area ABba
2. Adiabatic reversible expansion B to C	0	$w_2 = nR(T_1 - T_2)/(1-\gamma)$ = area BCcb
3. Isothermal reversible compression C to D	$-q_1$	$w_3 = -q_1 = -nRT_1 \ln(p_4/p_3)$ = − area DCcd
4. Adiabatic reversible compression D to A	0	$w_4 = -nR(T_1 - T_2)/(1-\gamma)$ = − area ADda
Total for cycle	$q_2 - q_1$	$nRT_2 \ln(p_1/p_2) - nRT_1 \ln(p_4/p_3)$ = area ABCD

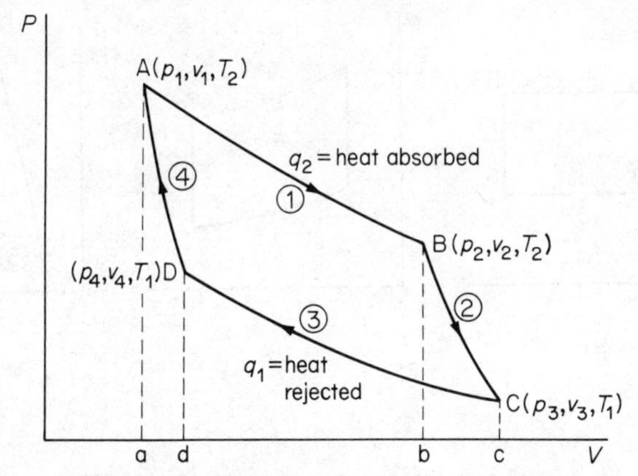

Figure C.1 Typical Carnot cycle for an ideal gas.

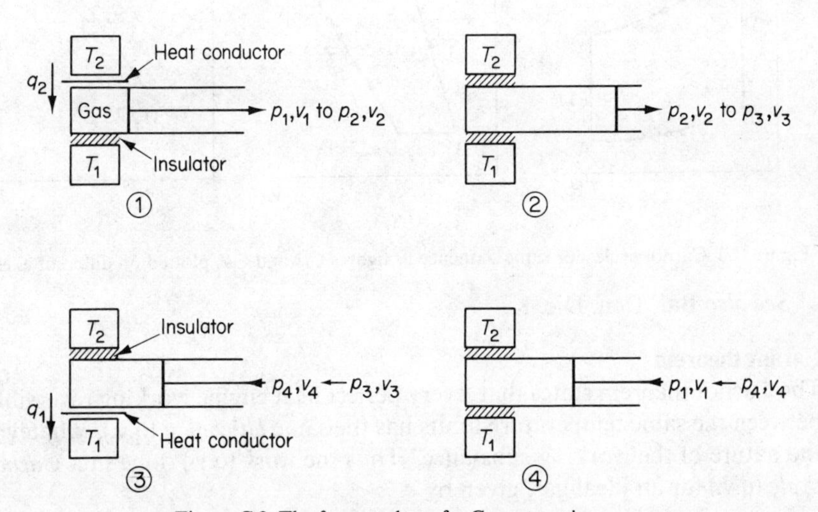

Figure C.2 The four strokes of a Carnot engine.

thus the *entropy* (q.v.) change in a reversible Carnot cycle is zero. Such a sequence of operations can be represented diagrammatically as in figures C.1 and C.2. The sequence of operations may be plotted with alternative axes (figure C.3).

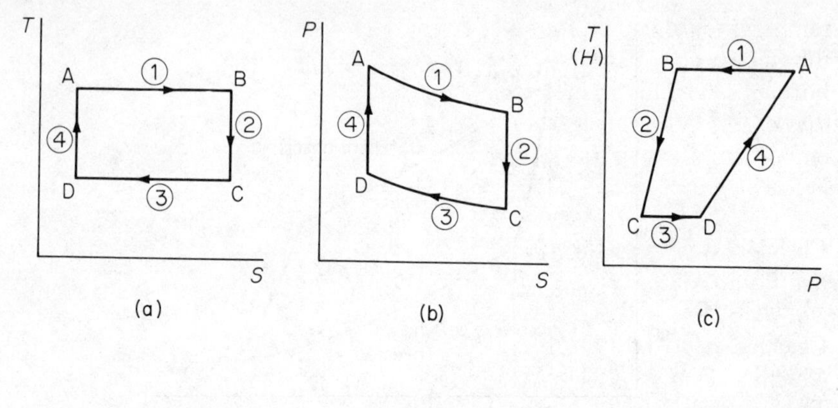

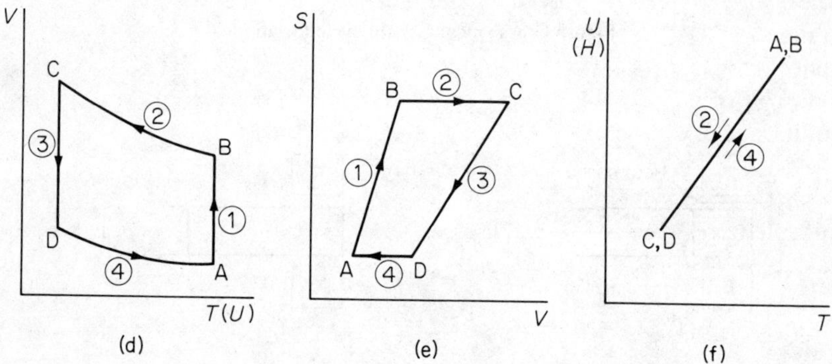

Figure C.3 Carnot cycles for same sequence as figures C.1 and C.2, plotted on different axes.

See also Bar, Den, Dic, K.

Carnot theorem

The Carnot theorem states that 'every perfect heat engine working reversibly between the same temperature limits has the same *efficiency* (q.v.), whatever the nature of the working substance.' Thus the *work* (q.v.) done in a *Carnot cycle* (q.v.) for an ideal gas, given by

$$w_{\max} = q_2(T_2 - T_1)/T_2$$

or, for a cycle working between T and $T - \mathrm{d}T$,

$$\text{đ}w = q(\mathrm{d}T/T)$$

is constant irrespective of the nature of the working substance. If a working substance could be found for which the efficiency between two temperatures was greater than that for any other working substance between the same temperatures, then, by coupling up the engines it would be possible, without the application of any outside effort, to transfer heat from a colder to a hotter body. This is in direct contradiction of the *second law of thermodynamics* (q.v.). The Carnot theorem thus provides a working equation embodying the principles of the second law.

See also Bar, Den, Dic, K.

Chemical equilibrium

See Equilibrium.

Chemical potential

Chemical potential, μ (units: J mol^{-1}), is a *partial molar quantity* (q.v.). The chemical potential of a component in a system corresponds to the change in the total *free energy* (q.v.) at constant T and P when 1 mol of that component is added to an infinite amount of the system.

For a general definition of chemical potential, the free energy of a multicomponent system is

$$G = f(T, P, n_A, n_B \ldots n_i)$$

and, hence, the over-all free energy change accompanying changes in all variables is

$$\mathrm{d}G = \left(\frac{\partial G}{\partial T}\right)_{P,n_j} \mathrm{d}T + \left(\frac{\partial G}{\partial P}\right)_{T,n_j} \mathrm{d}P + \left(\frac{\partial G}{\partial n_A}\right)_{P,T,n_j} \mathrm{d}n_A + \ldots \left(\frac{\partial G}{\partial n_i}\right)_{P,T,n_j} \mathrm{d}n_i$$

$$= -S\mathrm{d}T + V\mathrm{d}P + \mu_A \mathrm{d}n_A + \mu_B \mathrm{d}n_B + \ldots + \mu_i \mathrm{d}n_i$$

where the chemical potential of the ith component is

$$\mu_i = \left(\frac{\partial G}{\partial n_i}\right)_{T,P,n_j}$$

n_j is the number of moles of the remaining components.

The chemical potential, which may also be expressed in the equivalent forms,

$$\mu_i = \left(\frac{\partial G}{\partial n_i}\right)_{T,P,n_j} = \left(\frac{\partial U}{\partial n_i}\right)_{S,V,n_j} = \left(\frac{\partial H}{\partial n_i}\right)_{S,P,n_j}$$

$$= \left(\frac{\partial A}{\partial n_i}\right)_{T,V,n_j} = -T\left(\frac{\partial S}{\partial n_i}\right)_{U,V,n_j} = -T\left(\frac{\partial S}{\partial n_i}\right)_{H,P,n_j}$$

is not derived solely from one particular thermodynamic function such as G.

A change in the chemical potential can be used to calculate changes in G, U, H, A and S, depending on the constraints imposed on the process. At constant T and P, μ_i is closely identified with the Gibbs free energy per mole: i.e. $\mu_i = G_i/n_i = \bar{G}_i$.

(1) For a perfect gas. Since

$$\left(\frac{\partial G}{\partial P}\right)_T = V$$

it follows for an isothermal change that

$$\mathrm{d}G = V\mathrm{d}P$$

For 1 mole of a perfect gas, $G = \mu$ and $V = RT/P$; therefore

$$\mathrm{d}\mu = RT\ \mathrm{d}\ \ln P$$

which, on integration between two states 1 and 2, gives

$$\mu_2 = \mu_1 + RT\ \ln(P_2/P_1)$$

If state 1 represents a reference state (denoted $^\ominus$), the equation becomes

$$\mu = \mu^\ominus + RT\ \ln(P/P^\ominus)$$

The choice of the reference state is quite arbitrary, but it is convenient to use the standard pressure of 1 atm (101 325 N m^{-2}), so that

$$\mu = \mu^\ominus + RT\ \ln(P/\mathrm{atm})$$

where $\mu^\ominus$, the standard chemical potential at 1 atm pressure, depends on the temperature. The term $RT\ \ln P$ is a measure of the free energy difference between vapour at pressure P and at 101 325 N m^{-2}. The term $\mu^\ominus$ is the standard free energy of formation ($G_\mathrm{f}^\ominus$ or $\Delta G_\mathrm{f}^\ominus$) at the temperature in question. The term $\ln P$ refers to the logarithm of the numerical part only of P; the term $RT\ \ln(\mathrm{atm})$ is automatically compensated by a term, $-RT\ \ln(\mathrm{atm})$, in $\mu^\ominus$ which is not mentioned explicitly. If P is expressed in N m^{-2}, however, then it is necessary to divide this by 101 325 (i.e. add $-RT\ \ln 101\ 325$); thus the chemical potential of HCl vapour at 298 K and 10^5 N m^{-2} is given by

$$\begin{aligned}\mu(\mathrm{HCl}) &= \mu^\ominus(\mathrm{HCl}) + RT\ \ln P \\ &= -95.27\ + 8.314 \times 298 \times 2.303 \times 10^{-3}\ \log\frac{10^5}{101\ 325} \\ &= -95.24\ \mathrm{kJ\ mol^{-1}}\end{aligned}$$

(2) For the components of a perfect gas mixture,

$$d\mu_i = V_i \, dp = RT \, d \ln p_i$$

whence, by similar reasoning,

$$\mu_i = \mu_i^\ominus + RT \ln(p_i/\text{atm})$$

where $\mu_i^\ominus$ is the chemical potential of the *i*th component when its partial pressure is 1 atm.

(3) For real gases, which do not obey the gas laws, the fugacity, P^* or p^*, is introduced to make the ideal equations accurate; the 'activity coefficient' is defined as the ratio of fugacity to pressure, i.e. $P^* = fP$; thus, for a single imperfect gas,

$$\mu = \mu^\ominus + RT \ln P^* = \mu^\ominus + RT \ln fP$$

and, for a component of an imperfect gas mixture,

$$\mu_i = \mu_i^\ominus + RT \ln p_i^* = \mu_i^\ominus + RT \ln f_i p_i$$

(4) For a phase equilibrium. If the components in several phases are in equilibrium at fixed T and P, the transfer of dn_i mole of any component from one phase to another does not change the total free energy of the system, i.e.

$$\Delta G = \left(\sum_i \mu_i \, dn_i\right)_\alpha + \left(\sum_i \mu_i \, dn_i\right)_\beta + \ldots = 0$$

where α and β are the phases. The only solution of this equation, subject to the restriction that there is no loss of material from the system, i.e.

$$\left(\sum_i dn_i\right)_\alpha + \left(\sum_i dn_i\right)_\beta + \ldots = 0$$

is that $(\mu_i)_\alpha = (\mu_i)_\beta = \ldots$

Thus the chemical potential of a component is the same in all phases in equilibrium at a fixed temperature. (See also *phase rule*.)

(5) For the liquid–vapour equilibrium of one component. μ_g changes appreciably with P, but μ_l does not (figure C.4). Where the μ_g/T and μ_l/T curves intersect, the chemical potentials are equal and, hence, for a given pressure, there is only one temperature at which equilibrium is possible. For an ideal system,

$$\mu_l = \mu_g = \mu^\ominus + RT \ln P$$

and, for a non-ideal system,

$$\mu_l = \mu_g = \mu^\ominus + RT \ln fP$$

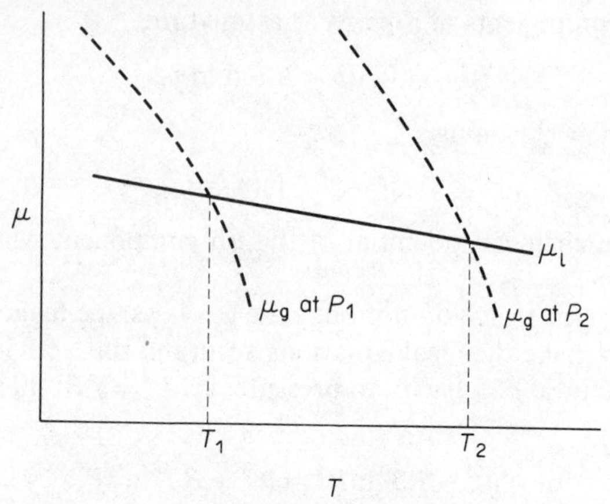

Figure C.4 Variation of the chemical potential of the gaseous and liquid states of a compound with temperature.

(6) For two components in two phases (liquid and vapour),

$$\mu_i(\text{soln}) = \mu_i(\text{vap}) = \mu_i^\ominus + RT \ln p_i$$

where p_i is the vapour pressure of the ith component over the solution (strictly the fugacity). For an ideal solution, *Raoult's law* (q.v.) is

$$p_i = p_i^\ominus x_i$$

where $p_i^\ominus$ is the vapour pressure of pure component i and x_i is its mole fraction in solution; hence,

$$\mu_i(\text{soln}) = \mu_i^\ominus + RT \ln p_i^\ominus + RT \ln x_i$$
$$= \mu_i^\ominus + RT \ln x_i$$

where $\mu_i^\ominus$, now containing the constant $RT \ln p_i^\ominus$, is the standard chemical potential when $x_i = 1$.

For a non-ideal system of two components,

$$\mu_A(\text{soln}) = \mu_A^\ominus + RT \ln f_A x_A \quad \text{and} \quad \mu_B(\text{soln}) = \mu_B^\ominus + RT \ln f_B x_B$$

or, in terms of molality,

$$\mu_A(\text{soln}) = \mu_A^\ominus + RT \ln \gamma_A m_A = \mu_A^\ominus + RT \ln a_A$$

where f_A, γ_A and f_B, γ_B are the *activity coefficients* (q.v.) of A and B, respectively, which approach unity as x_A and x_B, respectively, approach zero. The *activity* (q.v.) $a_A = \gamma_A m_A$.

(7) For an ion, the concept of chemical potential of one species of ion defined as

$$\mu_i = \left(\frac{\partial G}{\partial n_i}\right)_{T,P,n_{\mathrm{A}},n_j}$$

where i refers to one kind of ion, and A and j to the solvent and the other ion, respectively, is something of a mathematical fiction. Physically the operation implied by the equation cannot be achieved, since it means adding to the solution an ion of one kind only. The concept is, however, of use, and μ_i is defined as

$$\mu_i = \mu_i^\ominus + RT \ln a_i$$

where $\mu_i^\ominus$ is the standard chemical potential of the ion when $a_i = 1$. Thus in a system comprising AgCl(s), H_2O(l) which contains Ag^+, H^+ and Cl^-:

$$\begin{aligned}
&\mu(\mathrm{AgCl}) = \mu^\ominus(\mathrm{AgCl}) = -109.72\ \mathrm{kJ\ mol^{-1}}\\
&\mu(\mathrm{H_2O}) = \mu^\ominus(\mathrm{H_2O}) = -237.19\ \mathrm{kJ\ mol^{-1}}\\
&\mu(\mathrm{Ag^+}) = \mu^\ominus(\mathrm{Ag^+}) + RT \ln a(\mathrm{Ag^+}) = 77.1 + RT \ln a(\mathrm{Ag^+})\\
&\mu(\mathrm{Cl^-}) = \mu^\ominus(\mathrm{Cl^-}) + RT \ln a(\mathrm{Cl^-}) = 131.17 + RT \ln a(\mathrm{Cl^-})\\
&\mu(\mathrm{H^+}) = RT \ln a(\mathrm{H^+}) \qquad (\mu^\ominus(\mathrm{H^+}) = 0)
\end{aligned}$$

For dilute solutions, the chemical potential can be written

$$\dot{\mu}_i = \mu_i^\ominus + RT \ln (c_i/\mathrm{mol\ dm^{-3}})$$

In the term $\ln c_i$, only the numerical part of c_i is required, since it is automatically compensated by a term $-RT \ln(\mathrm{mol\ dm^{-3}})$ in $\mu_i^\ominus$.

(8) For an electrolyte, the chemical potential can be regarded as the sum of the separate chemical potentials of the individual ions. For an electrolyte, B,

$$G = \nu_1\mu_1 + \nu_2\mu_2 + RT \ln a_{\mathrm{B}}$$

and

$$G^\ominus = \nu_1\mu_1^\ominus + \nu_2\mu_2^\ominus$$

Thus

$$a_{\mathrm{B}} = a_1^{\nu_1}\, a_2^{\nu_2}$$

(9) For any ideal system, the chemical potential in terms of the *partition function* (q.v.), Q, is given by

$$\dot{\mu} = -kT \ln Q + kT \ln N_{\mathrm{A}}$$

where $\dot{\mu}$ is the chemical potential per molecule $= \mu/N_{\mathrm{A}}$.

Table C.2. Change of chemical potential for a perfect gas from the standard state at 298 K

Pressure/atm	Pressure/N m^{-2}	$(\mu - \mu^{\ominus})$/kJ
0.01	1 013.25	−11.42
0.05	5 066.25	− 7.40
0.1	10 132.5	− 5.69
0.5	50 662.25	−1.72
1.0	101 325	0
2	202 650	+ 1.72
10	1 013 250	+ 5.69
20	2 026 500	+ 7.40
100	10 132 500	+11.42

See also Dic, K, R & S, Wal, War, Wy.

Clausius–Clapeyron equation

The Clausius–Clapeyron equation relates the temperature dependence of the *vapour pressure* (q.v.) to the *enthalpy* (q.v.) and volume changes in a phase change. For the phase change,

$$\text{I} \rightleftharpoons \text{II}$$

(Form existing below *T*(transition)) (Form existing above *T*(transition))

$$\frac{dP}{dT} = \frac{\Delta H}{T\Delta V} = \frac{L_{tr}}{T(V_{II} - V_{I})}$$

The equation may be obtained from the fact that, when two phases are in equilibrium, the *free energy* (q.v.) of the substance is the same in both phases.

For the liquid–vapour equilibrium, the equation becomes

$$\frac{dP}{dT} = \frac{L_e}{T(V_g - V_l)}$$

where V_g and V_l are the volumes of the same amount of substance (e.g. molar volumes), in the vapour and liquid states, respectively, and L_e is the molar heat of vaporisation. Since $V_g \gg V_l$ and assuming ideal behaviour of the vapour, the equation becomes

$$\frac{d \ln P}{dT} = \frac{L_e}{RT^2}$$

Integration over small temperature ranges (L_e assumed constant) gives

$$\ln P = L_e/RT + C$$

or

$$\ln \frac{P_2}{P_1} = -\frac{L_e}{R}\left\{\frac{1}{T_2} - \frac{1}{T_1}\right\}$$

or

$$P = A \exp(-L_e/RT)$$

When the v.p. is known with great precision, or over a wide range of temperatures, the curvature due to the variation of L_e with T becomes apparent.

Similar considerations apply to the solid–vapour equilibrium, in which L_s replaces L_e.

The simple integrated form of the Clausius–Clapeyron equation may be used: (a) to determine the b.p. of a liquid, which decomposes at normal pressures, from measurements of b.p. at reduced pressure and a knowledge of L_e, and (b) to determine L_e (L_s) from a study of the variation of the v.p. of the liquid (solid) with temperature, using the plot of log P against T^{-1} (figure C.5).

For solid–liquid equilibria, the specific volumes of solid and liquid are approximately the same, so neither can be neglected with respect to the other. The equation is usually used in the form

$$\frac{dT}{dP} = \frac{T(V_l - V_s)}{L_f}$$

which shows how the m.p. or f.p. varies with the applied pressure.

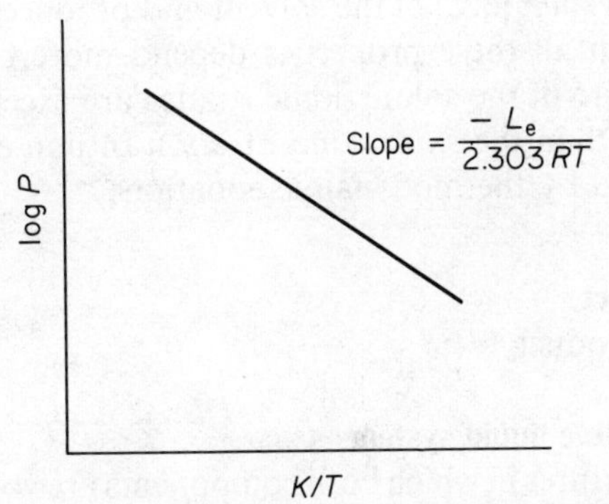

Figure C.5 Variation of vapour pressure of a liquid with temperature.

Two cases arise:

(1) $V_1 > V_s$, i. e. $\rho_s > \rho_1$, $dT/dP > 0$; thus an increase in pressure causes a raising of m.p.

(2) $V_1 < V_s$, i.e. $\rho_s < \rho_1$, $dT/dP < 0$; thus an increase in pressure causes a lowering of m.p. (See *water system.*)

An increase in pressure favours the stability of the phase with the lower molar volume.

For the ice–water system at 273 K, the specific volumes of liquid water and ice are 1.0000 and 1.0907 $m^3\ kg^{-1}$, respectively, and $L_f = 334.9 \times 10^3\ J\ kg^{-1}$; hence,

$$\frac{dT}{dP} = \frac{273.15(1.0000 - 1.0907) \times 10^{-3}}{334.9 \times 10^3} = -7.392 \times 10^{-8}\ K\ N^{-1}\ m^2$$

i.e. there is a decrease of 0.0074 K for an increase in pressure of 1 atm.

See also One-component system; Sulphur system; Water system; and Bar, Dic, K, Wy.

Closed system
See System.

Colligative properties
Colligative properties are those which depend on the number of particles present rather than on the kind of particles. Thus a solute dissolved in a solvent lowers the *vapour pressure* (q.v.), lowers the *freezing point* (q.v.), raises the *boiling point* (q.v.) of the solvent and produces an *osmotic pressure* (q.v.). Changes in all these properties depend more on the concentration than on the nature of the solute (ionic solutes are exceptional in producing greater changes than the same concentration of non-electrolyte). All these effects are related by thermodynamic equations.

Common ion effect
See Solubility product.

Completely miscible liquid systems
Binary liquid mixtures in which both components are volatile and completely miscible in the liquid phase may be divided into two types.

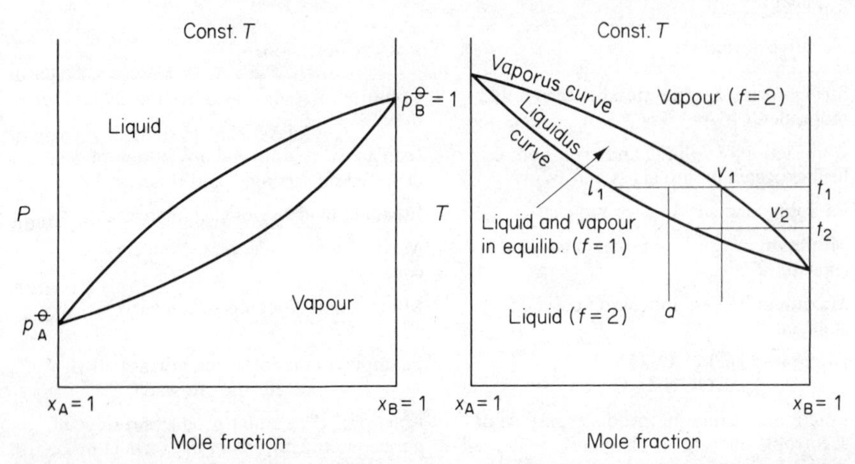

Figure C.6 Pressure–composition and temperature–composition curves for zeotropic systems.

(1) Zeotropic systems occur when the forces of attraction between like molecules are approximately equal to those between unlike molecules, so that there is little tendency for molecules of one component to alter the escaping tendency of the other component into the vapour, i.e. the two components are chemically similar (e.g. benzene and toluene, ethylene dibromide and propylene dibromide). In systems of this type, the v.p.–composition curve obeys *Raoult's law* (q.v.), i.e. the total pressure above the mixture is given in terms of the standard vapour pressures of the two components and their mole fractions in the liquid phase: $P = p_A^\ominus x_A + p_B^\ominus x_B$. In consequence, the v.p. and b.p. of all mixtures lie between the v.p. and b.p., respectively, of the two components. If the composition of the liquid and vapour in equilibrium at a given temperature and pressure are known, then it is possible to construct the liquidus and vaporus curves, respectively (figure C.6). Separation into the two pure components can be achieved by fractional *distillation* (q.v.), the ease of separation depending on the difference in b.p. of the two components; the greater the difference, the easier the separation.

When the system is entirely in the liquid or the vapour state, there are three *degrees of freedom* (q.v.), and pressure, temperature and composition can be independently specified. To describe such a system completely, a three-dimensional figure is required with pressure, temperature and compo-

Completely miscible liquid systems

Table C.3

Negative deviations	Positive deviations
Strong forces of attraction between unlike molecules, e.g. A – B > A – A	Cohesive forces between like molecules greater than between unlike, e.g. A – A > A – B
Contraction of volume and evolution of heat expected on mixing A and B	Increase of volume and absorption of heat expected on mixing A and B
Escaping tendency of each reduced	Escaping tendency of each increased
Minimum in v.p.–composition curve at constant *T*	Maximum in v.p.–composition curve at constant *T*
Maximum in *T*–composition curve at constant *P*	Minimum in *T*–composition curve at constant *P*
Examples: HCl/H_2O, HBr/H_2O, $HCOOH/H_2O$	Examples: ethanol/water, ethanol/ethyl acetate, dioxan/water
Fractional distillation produces residue of azeotropic mixture X; distillate is component present in excess in original mixture (figure C.7)	Fractional distillation produces residue of pure component present in excess in original mixture and distillate of azeotropic mixture Y (figure C.8)

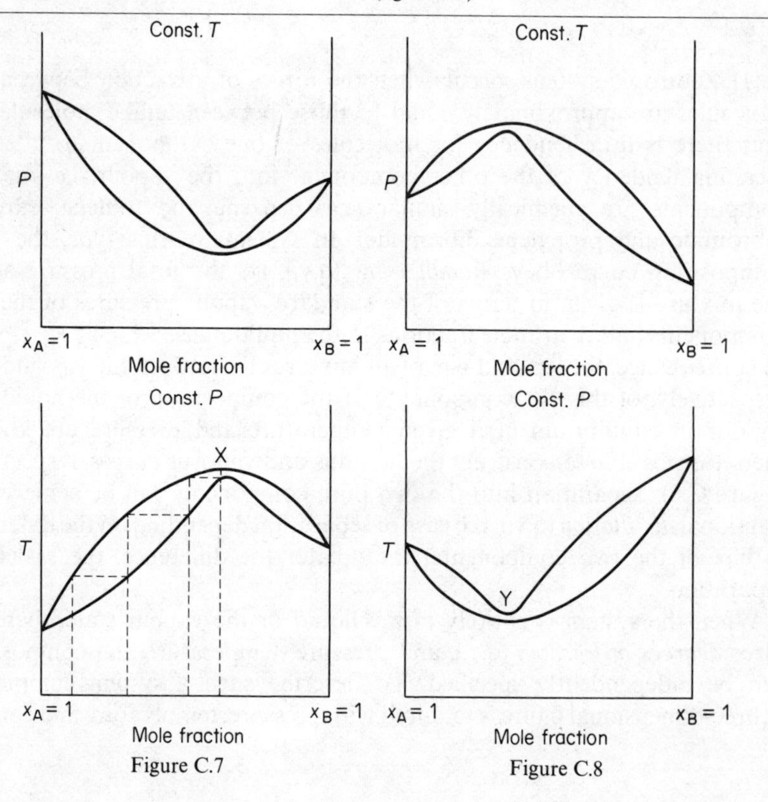

Figure C.7

Figure C.8

sition as the three coordinates. Such figures are difficult to construct and read, and so it is preferable to plot plane figures showing the dependence of P on composition at constant T (v.p.–composition curves) or of T on composition at constant P (b.p.–composition curves). Now the number of degrees of freedom is reduced to two, i.e. two degrees of freedom must be specified to define the system completely. Similarly, when liquid and vapour are in equilibrium, $f = 1$. If liquid of composition a is heated to t_1, liquid of composition l_1 is in equilibrium with vapour of composition v_1 (i.e. the vapour is richer in the more volatile component). Condensation of the vapour and further heating to t_2 gives enrichment of the vapour in B. Successive distillation and condensation eventually results in pure B; the residue becomes progressively richer in A until eventually pure A remains.

(2) Azeotropic systems, which occur when the components of real solutions are less similar, are divided into those showing positive and those showing negative deviations from Raoult's law (table C.3).

In some cases of deviation from ideality, the simple picture of cohesive forces is not adequate, e.g. strong positive deviations in aqueous solutions. Pure water is itself strongly associated, and the addition of a second component results in a partial depolymerisation of the water, leading to an increased v.p.

The azeotropic mixture, often known as a constant boiling mixture, is not a pure compound, since the composition and the b.p. vary with the external pressure, and the composition does not correspond to a simple formula. For example, HCl/H_2O (maximum b.p. mixture):

Pressure/mmHg	760	600	500	400
Conc. of HCl/wt%	20.22	20.64	20.92	21.24
Boiling point/°C	108.58		97.58	

and ethanol/water (minimum b.p. mixture):

Pressure/mmHg	95	760	2280
Conc. of EtOH/wt%	99.5	95.6	95.2

Azeotropic mixtures can be broken by the addition of a third component and redistilling, or by physical or chemical removal of one component.

The simplest method of plotting the liquidus and vaporus curves at constant P is to reflux a mixture of the two components until equilibrium is established; the temperature is recorded and small samples of the distillate and of the liquid in the flask in equilibrium are collected. Both samples are analysed, e.g. by measuring the refractive index or b.p. and comparing with a calibration curve. Thus the composition of liquid and vapour in equilibrium at a known temperature can be determined and plotted.

See also Bo, F & J, G, J & P.

Component

The number of components, c, of a system at equilibrium is the smallest number of independently variable constituents (i.e. molecular species), in terms of which the composition of each phase can be quantitatively defined (zero and negative quantities of the components are permissible). Molecular species must be chosen from the constituents which are present when the system is in a state of equilibrium and which take part in the equilibrium. For the ice–water–water vapour system, each phase is made up of one molecular species H_2O; hence, $c = 1$. In the equilibrium

$$CaCO_3(s) \rightleftharpoons CaO(s) + CO_2(g)$$

$c = 2$, which may be $CaCO_3$ and CaO; $CaCO_3$ and CO_2 or CaO and CO_2. For the dissociation

$$NH_4Cl(s) \rightleftharpoons NH_3(g) + HCl(g)$$

$c = 1$, since the gas phase is made by vaporising the solid, and the concentrations of NH_3 and HCl are equal; however, if excess $NH_3(g)$ is pumped into the system containing the ammonium chloride equilibrium, then it becomes a two-component system.

While the identity of the components is subject to some degree of choice, the number of components for a given system is definitely fixed. (N.B. Its value may change with change of conditions such as temperature.)

See also Phase rule; and Bo, F & J.

Concentration units

The concentration of a solution may be expressed in many different ways, all involving the amount of solute and the amount of either solvent or solution. The various units can be divided into two groups: (1) those measuring concentration as weight of solute per unit volume of solution, e.g. g solute per dm^3 solution, mole of solute per dm^3 of solution; (2) those measuring concentration as weight of solute per unit weight of solution, e.g. weight per cent, g of solute per kg of solvent, mole of solute per kg of solvent (molality), mole fraction and mole per cent.

Group (2) units, which refer solely to weight, have two definite advantages over the group (1) units, which involve volumes; firstly, the weights of the constituents are additive, while their volumes rarely are; secondly, weight concentrations are independent of temperature.

Interconversion of units: consider a w wt% solution of compound A (relative molecular mass $M_r(\mathrm{A})$) in water $[M_r(\mathrm{H_2O}) = 18]$ at 20°C; density of solution $= \rho$ kg m^{-3}. This solution contains w g of A to $(100 - w)$ g water

$$\text{Concentration} = \frac{w}{(100 - w)} \times 1000 \text{ g of A per kg water}$$

Hence,

$$\text{molality} = \frac{w}{(100 - w)} \times \frac{1000}{M_r(\mathrm{A})} \text{ mol kg}^{-1}$$

The *mole fraction* (q.v.) is calculated from the number of moles of each component in 100 g of the solution:

$$n_{\mathrm{A}} = w/M_r(\mathrm{A}); \qquad n(\mathrm{H_2O}) = (100 - w)/18$$

$$x_{\mathrm{A}} = \frac{w/M_r(\mathrm{A})}{w/M_r(\mathrm{A}) + (100 - w)/18}$$

$$\text{Mole \% of A} = \frac{w/M_r(\mathrm{A})}{w/M_r(\mathrm{A}) + (100 - w)/18} \times 100$$

The value of the density is required to convert wt/wt solutions to wt/vol solutions or vice versa. 1 m^3 of solution weighs 1000 ρ kg, and, since w% of this is A, the solution therefore contains

$$\frac{1000\,\rho\,w}{100} \text{ kg of A per m}^3 \text{ of solution}$$

$$\text{Hence, concentration} = \frac{w}{100} \times \frac{10^3\,\rho}{M_r(\mathrm{A})} \text{ mol m}^{-3}$$

$$= \frac{w}{100} \times \frac{\rho}{M_r(\mathrm{A})} \text{ mol dm}^{-3}$$

Condensed system

A condensed system is a system composed only of liquid and/or solid phases.

See also Phase rule; Three-component system; Two-component system.

Congruent melting point

A compound is said to have a congruent *melting point* (q.v.) if the solid substance can exist in equilibrium with liquid of the same composition. The m.p. of the compound may be above or below the m.p. of the two components in a *two-component condensed system* (q.v.).

Conjugate solutions
When two *partially miscible liquids* (q.v.) are shaken together, two liquid layers may be formed. These layers in equilibrium at a fixed temperature and pressure are known as conjugate solutions; they are actually saturated solutions of one component in the other. The compositions of the two solutions are joined by a *tie line* (q.v.) on a phase diagram and, at the fixed temperature, their composition remains unchanged although the relative amounts of the two phases may change.

Conjugate ternary solutions
See Three-component system.

Conservation of energy
See First law of thermodynamics.

Consolute temperature
See Critical solution temperature; Partially miscible liquids

Constant heat summation
See Hess's law of constant heat summation.

Cooling curves
See Thermal analysis.

Cottrell method
In the Cottrell method for the determination of the elevation of the *boiling point* (q.v.), the elevation of b.p. is measured under conditions such that superheating is eliminated. The boiling liquid pumps itself over the bulb of a *Beckmann thermometer* (q.v.) (figure C.9), so that a thin layer of solution, which readily comes to equilibrium with the vapour at atmospheric pressure, covers the bulb of the thermometer. This then registers the true b.p., eliminating errors due to both superheating and hydrostatic pressure. The whole apparatus must be adequately screened from draughts and the platinum wire heated by a small flame. From the measured value of ΔT_e, the relative molecular mass of the solute can be determined.

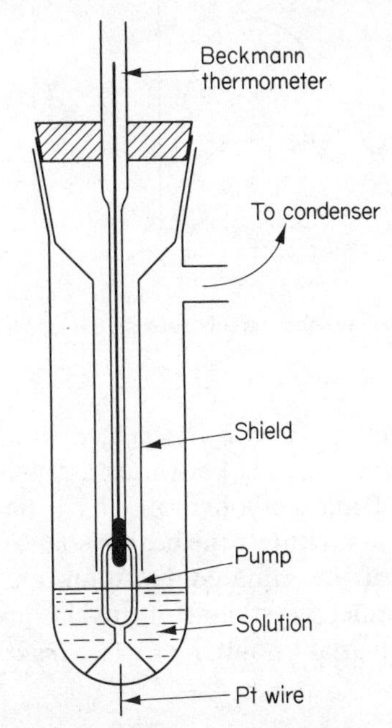

Figure C.9 Cottrell apparatus showing pump.

This pumping principle is used in the commercially available semi-microebulliometers; the detecting device is a *thermistor* (q.v.).

See also J & P.

Countercurrent distribution

See Distribution law.

Critical solution temperature

For *partially miscible liquids* (q.v.) in which the mutual solubility increases (decreases) with increasing temperature, the upper (lower) critical solution temperature or consolute temperature, (C) is the temperature above (below) which the liquids are miscible in all proportions (figure C.10). The critical solution temperature represents a fixed temperature and composition, and depends on the nature of the components; it is sensitive to changes of pressure and also to the presence of impurities.

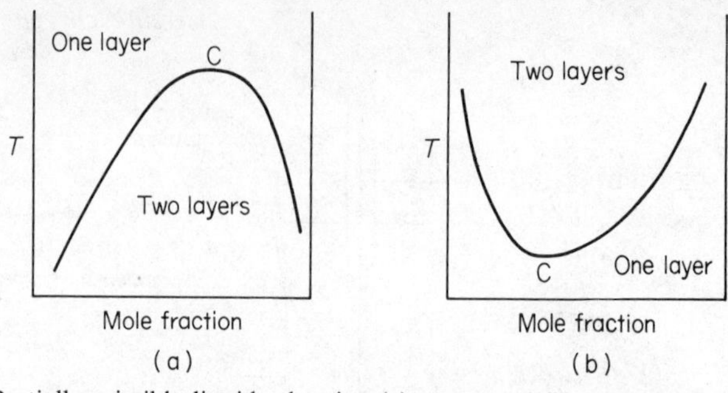

Figure C.10 Partially miscible liquids showing (a) upper and (b) lower critical temperature.

Cryohydrate

In a *two-component condensed system* (q.v.) in which water is one component, the *eutectic mixture* (q.v.) is known as a cryohydrate (figure C.11). Despite the constancy of m.p., cryohydrates have the properties not of chemical compounds but of mixtures; the heat of solution is the same as for a mixture of ice and salt of the same composition, and microscopical examination reveals two distinct crystalline forms in juxtaposition. The cryohydric point (C) is an invariant point, $f' = 0$ (see *invariant system*).

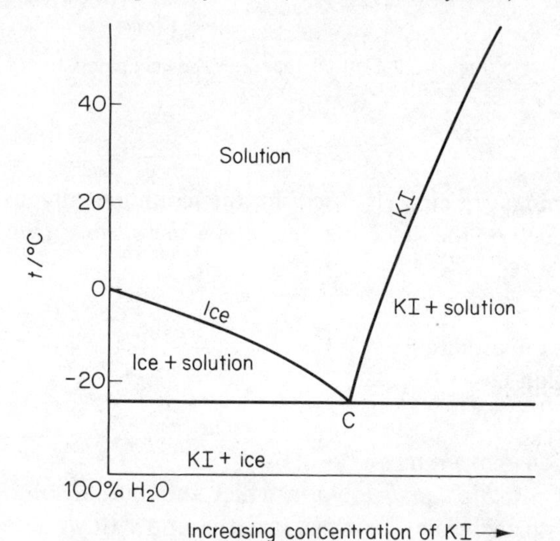

Figure C.11 System of salt (KI) and water.

Cryoscopic constant

The cryoscopic constant, k_c (units: K kg mol^{-1}), or the molal depression of the *freezing point* (q.v.) of a solvent A, is defined by the equation

$$k_c = \frac{RT_0^2}{n_A L_f} = \frac{RT_0^2}{l_f}$$

where n_A is the number of moles of solvent in 1 kg solvent. Where a choice of solvent is possible, it is best to use that with the smallest value of l_f; in this way, the largest change in temperature for a given concentration of solute is obtained.

Cycle

A cycle or cyclic process is a process which leads from a system in a given state through a sequence of changes back to the system in its original state; e.g., at constant pressure, the temperature of a system may be increased by performing frictional work and then lowered to its original value by the removal of heat. For any cycle and any *state function* (q.v.), X,

$$\oint \mathrm{d}X = \Delta X = 0$$

In the isothermal cyclic expansion (figure C.12) of a gas from state I to state II followed by compression to the initial state I, *work* (q.v.) is done by the gas on the surroundings; this is paid for by supplying an equivalent amount of *heat* (q.v.) to make the cycle operate (e.g. in the steam engine), since

$$\Delta U = q - w = 0$$

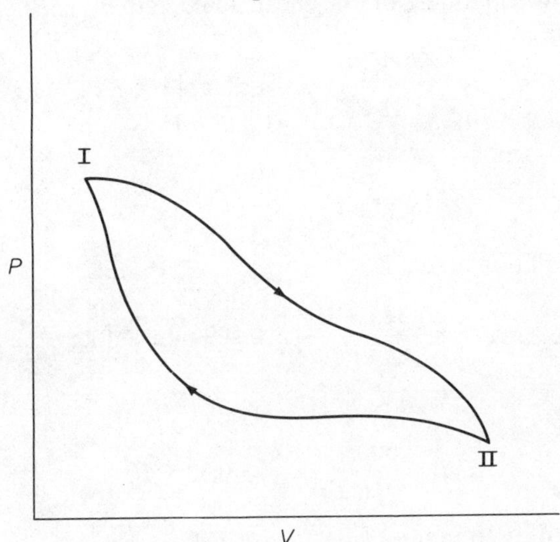

Figure C.12 Typical isothermal cycle for a perfect gas.

See also Carnot cycle.

D

Debye–Hückel activity equation

The Debye–Huckel activity equation is a theoretical equation which relates the *activity coefficient* (q.v.) of an ion in an electrolyte to the *ionic strength* (q.v.) of the solution. In the derivation of the equation, it is assumed that the only attractive forces between *ions*† are electrostatic and that the relative permittivity remains constant throughout the solution and right through the ionic atmosphere:

$$\log \gamma_i = -\frac{A\, z_i^2\, I^{1/2}}{1 + B\, \mathring{a}\, I^{1/2}}$$

where $\mathring{a}$ may be regarded as the effective diameter of the ion. The numerator gives the effect of long-range Coulomb forces and the denominator the modification due to short-range forces, calculated on the basis of a hard sphere model of the electrolyte. As $I \longrightarrow 0$, the equation reduces to the limiting equation:

$$\log \gamma_i = -A\, z_i^2\, I^{1/2}$$

The activity coefficient of a single ionic species cannot be measured. The corresponding equations for the mean ionic activity coefficient $\gamma_\pm$ of the electrolyte are:

$$\log \gamma_\pm = -\frac{A|z_+\, z_-|\, I^{1/2}}{1 + B\, \mathring{a}\, I^{1/2}}$$

and

$$\log \gamma_\pm = -A|z_+\, z_-|\, I^{1/2}$$

In aqueous solution at 298 K,

$$A = \frac{1.825 \times 10^6}{(\varepsilon T)^{3/2}} = 0.509\ \mathrm{mol}^{-1/2}\ \mathrm{kg}^{1/2}$$

and

$$B = \frac{50.29 \times 10^8}{(\varepsilon T)^{1/2}} = 0.33 \times 10^{10}$$

(where $\mathring{a}$ is expressed in m).

The equations predict that the limiting slope of the $\log \gamma_\pm - I^{1/2}$ curve should be negative, proportional to the valence product of the electrolyte and inversely proportional to $\varepsilon^{3/2}$ and to $T^{3/2}$ (through the constant A). The B term in the denominator indicates that all the activity coefficient curves should diverge upwards from the limiting law as the concentration increases. Experimental results for solutions of sufficiently low I are in good agreement with the theoretical predictions.

See also tables A.VII and A.VIII (pp. 255, 256); and G, R & S.

Degree of freedom

Degree of freedom or variance, f, is the number of intensive state variables that can be independently varied without changing the number of phases present in equilibrium, e.g. the state of a given amount of gas is specified by any two of the variables temperature, pressure and density; thus $f = 2$ (bivariant system). For the water–water vapour system, only one variable is necessary to define the system, since at a given temperature the v.p. of water is fixed, $f = 1$ (univariant system); the system in which ice, water and water vapour are in equilibrium is invariant, $f = 0$.

See also Phase rule; and Bo, F & J.

Deliquescence

Deliquescence occurs when the partial water v.p. exceeds that of the aqueous saturated salt solution, and the surface of the *salt hydrate* (q.v.) becomes covered with a layer of saturated solution. The saturation v.p. of water at room temperature is about 15 mmHg, so highly soluble salts may form solutions having v.p. below 10 mmHg; thus both the salt and the saturated solution will absorb water from the atmosphere. Salts forming hydrates with very low v.p. will be deliquescent: e.g., at 20 °C for $CaCl_2 \cdot 6H_2O$, the v.p. of the hydrate = 2.5 mmHg and the v.p. of the saturated solution = 7.5 mmHg.

See also Efflorescence.

Depression of freezing point

See Freezing point.

Depression of vapour pressure

See Raoult's law; Vapour pressure.

Dielectric constant

The dielectric constant or relative permittivity, ε_r (a dimensionless factor), for a given dielectric medium is the ratio of the capacitance, C, of a capacitor with the region between the plates filled with the material to the capacitance,

C_0, of the same capacitor when the region between the plates is a vacuum, i.e.

$$\varepsilon_r = C/C_0$$

ε_r is familiar as the proportionality constant in Coulomb's law for the force, F, between two charges, Q and Q', separated by a distance, r, in a uniform medium:

$$F = \frac{Q\,Q'}{r^2\,\varepsilon_r}$$

Values of ε_r for liquids vary in the range 1–120; e.g., at 298 K, cyclohexane 2.015, benzene 2.274, chlorobenzene 5.621, methanol 32.63, nitrobenzene 34.82, water 78.54.

Differential thermal analysis

Differential thermal analysis (DTA) is widely used to observe phase changes and to determine the heat changes involved in them. An inert material and the sample under study are heated simultaneously in the same metal block; a thermocouple in each records the temperature difference between them. As long as there is no phase change, there will be no temperature difference; when the sample undergoes a phase change in which heat is absorbed, the temperature of the sample will be less than that of the reference material. Once the phase change is complete (provided that the weight of the block is much greater than the weight of the sample), the temperature difference will rapidly decrease, as shown in figure D.1.

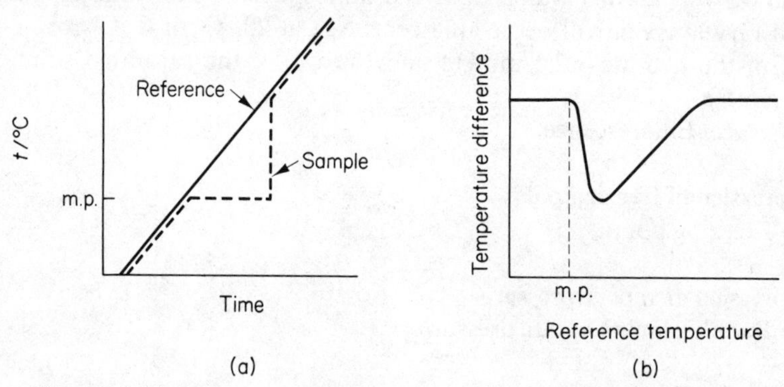

Figure D.1 (a) Comparative heating curves. (b) Differential temperature curve.

Disorder

See Entropy.

Dissociation constant

The dissociation constant or ionisation constant, K_a (units: mol dm^{-3}), for a weak acid, HA,

$$HA + H_2O \rightleftharpoons H_3O^+ + A^-$$

is given by

$$K_a = \frac{a(H_3O^+)\,a(A^-)}{a(HA)} = \frac{c(H_3O^+)\,c(A^-)}{c(HA)} \times \frac{\gamma(H_3O^+)\,\gamma(A^-)}{\gamma(HA)}$$
$$= K_c\,\gamma(H_3O^+)\,\gamma(A^-) = K_c\,\gamma_{\pm}^2 \qquad \text{(D.1)}$$

and $pK_a = -\log K_a$.

Weak bases can be treated in a similar manner.

Determination of dissociation constants

(1) Titration curves (figure D.2) (*electrometric titrations*†) only give the classical dissociation constant, K_c; a correction may be applied using the *Debye–Hückel activity equation* (q.v.) to give K_a. After the addition of t cm^3 of titrant, the concentration of free acid is proportional to $(T - t)$, where T is the volume of titrant at the equivalence point, and the concentration of salt is proportional to t. Hence, using the Henderson equation,

$$pH = pK_c + \log\frac{c(A^-)}{c(HA)} = pK_c + \log\frac{t}{(T-t)} \qquad \text{(D.2)}$$

The graph of pH against $\log t/(T - t)$ is linear and of intercept pK_c (figure D.3). At the point of 'half titration', $t = T/2$; hence, $pH = pK_c$.

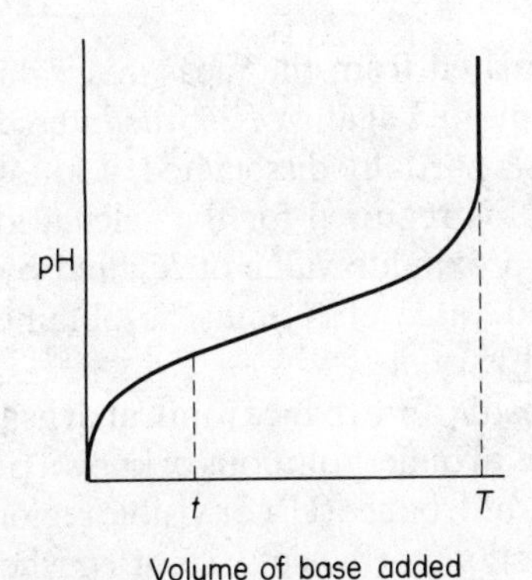

Figure D.2 Typical titration curve.

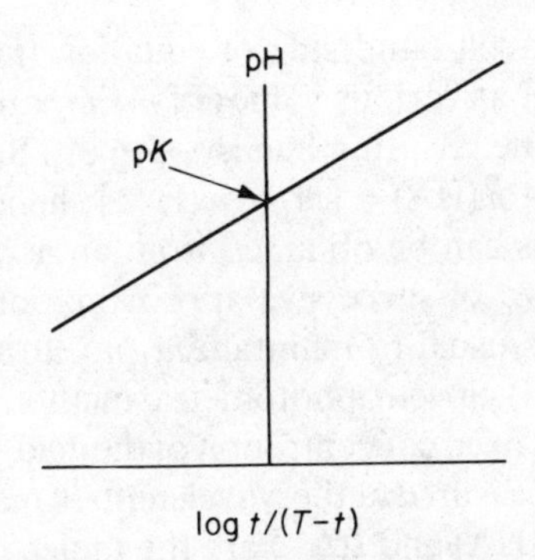

Figure D.3 Graph of pH against $\log t/(T - t)$.

(2) Conductometric method. From the measured *conductance*† of a weak acid at low concentrations, the degree of dissociation $\alpha(=\Lambda_c/\Lambda'_e)$ and $K_c(=\alpha^2 c/(1-\alpha))$ can be calculated. Λ'_c, the molar conductivity which the electrolyte would have if it were fully dissociated at concentration c, may be calculated using the Kohlrausch law of independent migration of ions

$$\Lambda'_c(\mathrm{HA}) = \Lambda_c(\mathrm{HCl}) + \Lambda_c(\mathrm{NaA}) - \Lambda_c(\mathrm{NaCl})$$

where the Λ_c values for the strong electrolytes all refer to the same concentration c. Since

$$\log K_c = \log K_a + 2\,A\,(\alpha\,c)^{1/2} \qquad \text{(D.3)}$$

the graph of $\log K_c$ against $(\alpha\,c)^{1/2}$ is linear and of intercept $\log K_a$.

(3) EMF method using *cells*† without *liquid junctions*†, e.g.

$$\ominus\ \mathrm{Pt}, \mathrm{H_2(g)}\,|\,\mathrm{HA}(m_1), \mathrm{NaA}(m_2), \mathrm{NaCl}(m_3)\,|\,\mathrm{AgCl(s)}, \mathrm{Ag}\ \oplus$$

for which the over-all cell reaction is

$$\tfrac{1}{2}\mathrm{H_2(g)} + \mathrm{AgCl(s)} \longrightarrow \mathrm{Ag(s)} + \mathrm{H^+} + \mathrm{Cl^-}$$

and the e.m.f. is

$$E = E^{\ominus}(\mathrm{AgCl, Ag, Cl^-}) - \frac{RT}{F}\ln a(\mathrm{H^+})\,a(\mathrm{Cl^-}) \qquad \text{(D.4)}$$

The hydrogen ions are provided by the dissociation of the weak acid; substituting for $a(\mathrm{H^+})$ from equation (D.1) in equation (D.4) gives

$$\frac{F[E-E^{\ominus}(\mathrm{AgCl, Ag, Cl^-})]}{2.303\,RT} + \log\frac{m(\mathrm{HA})\,m(\mathrm{Cl^-})}{m(\mathrm{A^-})} = -\log\frac{\gamma(\mathrm{HA})\,\gamma(\mathrm{Cl^-})}{\gamma(\mathrm{A^-})} - \log K_a \qquad \text{(D.5)}$$

The left-hand side of equation (D.5), calculated from the measured values of E at various values of m_1, m_2 and m_3, is plotted against $I^{1/2}$; the intercept of the resulting line is $-\log K_a$. Since HA is partially dissociated, it follows that $m(\mathrm{HA}) = [m_1 - m(\mathrm{H^+})]$; hence, $m(\mathrm{H^+})$ is required for the calculation. This can be obtained from an assumed approximate value of K_a, and by a series of successive approximations a good value of K_a may be obtained. It is usual to maintain m_1/m_2 at unity and vary m_3.

(4) Spectrophotometric method. The decadic absorbance (optical density) of a fixed concentration of the acid, in a series of buffer solutions of known pH, is measured at the wavelength of maximum absorbance (UV or visible region). If $\varepsilon(\mathrm{HA})$ and $\varepsilon(\mathrm{A^-})$ are the molecular extinction coefficients (i.e. A/cl, where A is the absorbance and l the path length) of HA and $\mathrm{A^-}$, respectively, and

ε/dm^2 mol^{-1} is the measured extinction coefficient, and c/mol dm^{-3} is the total acid concentration, then

$$\varepsilon c = \varepsilon(\mathrm{HA})\, c(\mathrm{HA}) + \varepsilon(\mathrm{A}^-)\, c(\mathrm{A}^-) \qquad (\mathrm{H_3O^+} \text{ does not absorb})$$

and

$$c = c(\mathrm{HA}) + c(\mathrm{A}^-)$$

Thus

$$\mathrm{pH} = \mathrm{p}K_\mathrm{a} + \log \frac{c(\mathrm{A}^-)}{c(\mathrm{HA})} + \log \gamma(\mathrm{A}^-)$$

$$= \mathrm{p}K_\mathrm{a} + \log \left\{\frac{\varepsilon(\mathrm{HA}) - \varepsilon}{\varepsilon - \varepsilon(\mathrm{A}^-)}\right\} - \frac{A\, z^2\, I^{1/2}}{1 + 1.25\, I^{1/2}}$$

Hence, the graph of pH against $\log [\varepsilon(\mathrm{HA}) - \varepsilon]/[\varepsilon - \varepsilon(\mathrm{A}^-)]$ is linear and of intercept $\mathrm{p}K_\mathrm{a} - A\, z^2\, I^{1/2}/(1 + 1.25\, I^{1/2})$.

A decrease in the relative permittivity (see *dielectric constant*) of the solvent causes an increase in the electrostatic forces between anions and cations and, hence, an increase in the formation of covalent bonds. This is accompanied by a decrease in the dissociation constant of a weak acid dissolved in it. For acetic acid at 298 K, K_a in water ($\varepsilon_\mathrm{r} = 78.5$) = 1.754×10^{-5} mol dm^{-3}, while in 82% dioxane ($\varepsilon_\mathrm{r} = 9.5$) $K_\mathrm{a} = 3.1 \times 10^{-11}$ mol dm^{-3}.

The dissociation constant of a weak acid varies with the temperature according to an equation of the type

$$2.303\, R \log K_\mathrm{a} = -A/T + C - DT$$

where A, C and D are constants. K_a passes through a maximum value at a temperature of $(A/D)^{1/2}$, which is about room temperature for most weak acids. The maximum value is given by

$$2.303\, R \log K_\mathrm{a} = C - 2\,(AD)^{1/2}$$

See also table A.X (p. 257); and J & P, K, M, R & S.

Distillation

Distillation is a process by which volatile components may be purified and separated from other components in a mixture. It is usually carried out under isobaric conditions (see *isobaric process*), and the temperature of the mixture is raised until the total v.p. equals the external pressure (usually atmospheric); under these conditions the vapour in equilibrium with the liquid is richer in the more volatile component. Distillation can also be

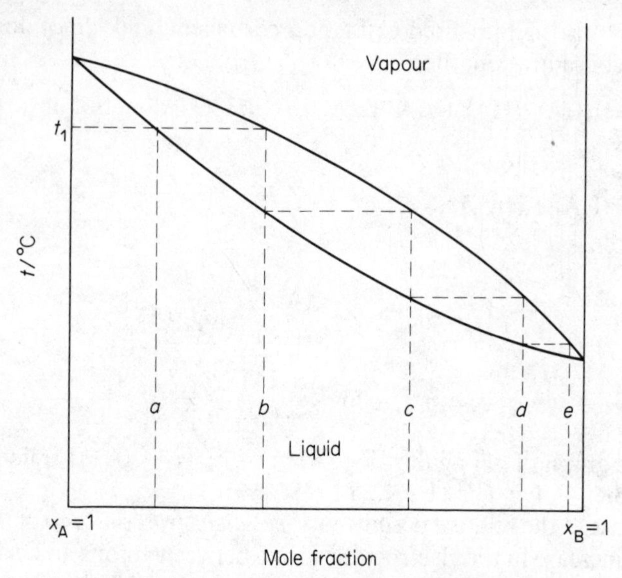

Figure D.4 Boiling point–composition diagram for near-ideal binary liquid mixture.

carried out under isothermal conditions (see *isothermal process*), in which case the external pressure is reduced until boiling proceeds and liquid and vapour are in equilibrium. Fractional distillation is the separation of the two components of a *binary liquid mixture* (q.v.) by repeated distillation and condensation, or by the use of a fractionating column.

If the near-ideal binary liquid mixture of composition a (figure D.4) is heated isobarically to temperature t_1, then the vapour in equilibrium with the liquid is richer in the more volatile component B. If this vapour is removed, then some separation has been achieved; appreciable separation by this one step is possible only when there is a large difference between the b.p.s of the two components. In the process of fractional distillation, this single step is repeated many times by condensing and boiling off again. The effect of this is to pass across the b.p.–concentration diagram from a–b–c–d–e. It is thus possible, in theory, to separate the mixture into the pure components A and B. The ease with which this separation is achieved for a *zeotropic mixture* (q.v.) depends on the relative b.p.s of the two components. Fractional distillation of azeotropic mixtures (see *azeotrope*) results in the separation into the pure component present in excess and the azeotropic mixture (either a maximum or minimum b.p. mixture).

Experimentally this fractionation process is achieved automatically by

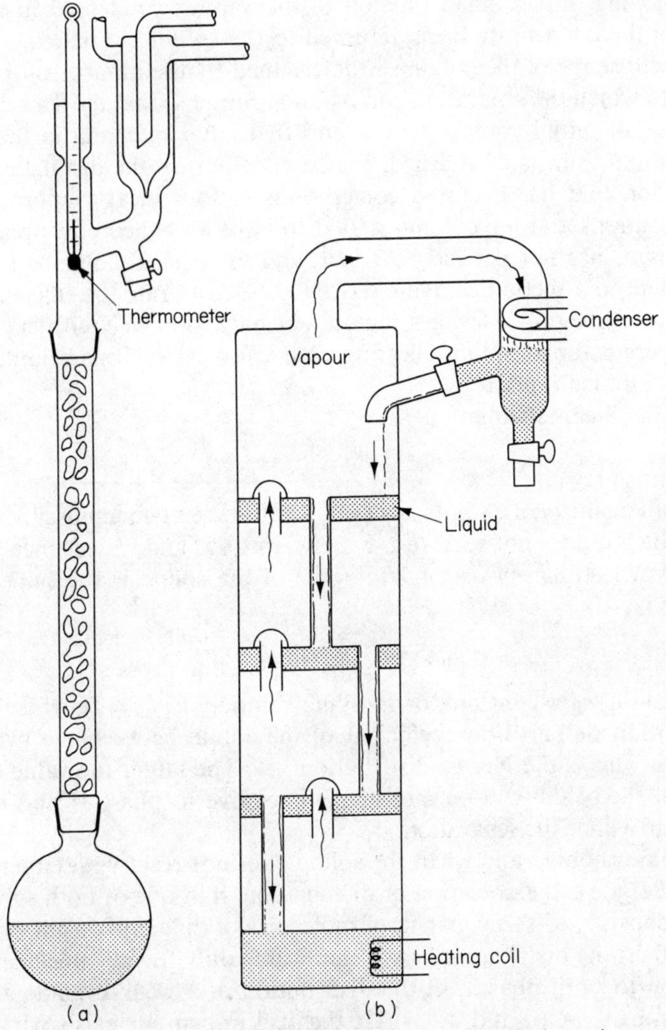

Figure D.5 Laboratory and industrial fractional distillation units.

using a fractionating column packed with glass beads, helices, etc., above the flask containing the mixture. The column head (figure D.5) is fitted with a thermometer and condenser for the distillate. The system of ascending vapour and returning liquid is maintained as near as possible at equilibrium by removing only a small fraction of the vapour condensed in the head, most of the condensate being returned to the column in reflux.

The efficiency of the column is determined by the number of theoretical plates to which the separation corresponds. Suppose that the flask contained a charge of initial composition *a*, and that, after refluxing in the column until equilibrium is established, the composition of the distillate is *e*. The separation that has occurred corresponds to four ideal evaporations and condensations and the column is said to have four theoretical plates.

Columns are not perfectly efficient, and for packed columns the height equivalent to a theoretical plate (HEPT) is used to rate the efficiency of the column. A tube only a few feet high can be packed to be equivalent to about 100 theoretical plates, thus effecting practically perfect separation of all but the most difficult mixtures.

See also Steam distillation.

Distribution law

If a small quantity of a solute is distributed between two immiscible solvents, with which it does not react (e.g. I_2 between CCl_4 and water), then at *equilibrium* (q.v.) the *chemical potential* (q.v.) of the solute is the same in both phases; i.e.

$$\mu_{\mathrm{I}} = \mu_{\mathrm{II}}; \quad \text{or} \quad RT \ln a_{\mathrm{I}} = RT \ln a_{\mathrm{II}}$$

whence $\ln a_{\mathrm{I}}/a_{\mathrm{II}}$ = constant, or in dilute solutions $c_{\mathrm{I}}/c_{\mathrm{II}} = D$, where D is the distribution or partition coefficient of the solute between the two phases I and II. This is the Nernst distribution law. The larger the value of D, the more of the solute will go into phase I relative to phase II and the more complete will be the separation.

The law is only valid when the solute does not react with either solvent, and undergoes no association or dissociation. If in one or both solvents the solute consists, to some extent, of molecules of different weight or composition, then the distribution law is applicable only to the molecular species common to both phases. In the distribution of a weak organic acid, HA, between benzene (I) and water (II), the acid in benzene exists partly as the dimer, while in water it is partly dissociated. The species common to both solvents is HA, and if α represents the fraction of the total amount of solute associated or dissociated, then

$$\frac{(1-\alpha_{\mathrm{I}})\, c_{\mathrm{I}}}{(1-\alpha_{\mathrm{II}})\, c_{\mathrm{II}}} = D$$

If solute B has a normal molecular mass in I but associates in II, i.e.

$$n\,\mathrm{B}\ (\text{in I}) \rightleftharpoons \mathrm{B}_n\ (\text{in II})$$

and if c_{II} is the total concentration of B in II and α is the fraction which has undergone association, then the equilibrium constant is

$$K = \frac{\alpha\, c_{\mathrm{II}}}{n\,(1-\alpha)^n\, c_{\mathrm{II}}^n}$$

or

$$(1-\alpha)\, c_{\mathrm{II}} = (\alpha\, c_{\mathrm{II}}/nK)^{1/n}$$

Application of the distribution law gives

$$D = \frac{c_{\mathrm{I}}}{(1-\alpha)\, c_{\mathrm{II}}} = \frac{c_{\mathrm{I}}}{(\alpha\, c_{\mathrm{II}}/nK)^{1/n}}$$

When molecules in II are almost completely associated, $\alpha \approx 1$ and $c_{\mathrm{I}}/c_{\mathrm{II}}^{1/n} =$ constant, since n, K and D are all constants. Thus, for a given system, the graph of log c_{I} against log c_{II} is linear and of slope $1/n$, giving an approximate value of n.

Like all *equilibrium constants* (q.v.), D varies with T and with concentration; hence, there is a need to correct for deviations from ideal behaviour by replacing concentrations with activities.

The distribution principle is of use in determining equilibrium constants (e.g. triiodide system, $I_2 + I^- \rightleftharpoons I_3^-$; weak acids) and the formulae of complex ions (e.g. $Cu(NH_3)_4^{2+}$) and in extraction processes including countercurrent distribution. When a solute is distributed between two solvents (without any complications), the weight of solute that can be removed in a series of extractions can be calculated. If a solution containing W g of solute in V_{I} cm^3 of solution is shaken successively n times with aliquots (each of V_{II} cm^3) of a second immiscible solvent until equilibrium is established, then the weight of solute remaining unextracted is

$$W_n = W\left(\frac{D\,V_{\mathrm{I}}}{D\,V_{\mathrm{I}} + V_{\mathrm{II}}}\right)^n$$

This equation may be used to estimate the number of extractions necessary to reduce W to a given value W_n. When a limited volume of solvent is available for extraction, greater extracting efficiency is obtained when this volume is used in a number of separate extractions than would be the case if it were used all at once (i.e. V_{II} small and n large).

See also G & S, J & P.

Donnan membrane equilibrium

Donnan membrane equilibrium concerns the distribution of simple electrolytes on the two sides of a membrane, freely permeable to these electrolytes but impermeable to colloidal ions, in the presence of a colloidal electrolyte on one side of the membrane. If an aqueous solution of the sodium salt of a protein Na_zR (R is a colloidal anion) of concentration c_1 is separated by such a membrane from an equal volume of NaCl of concentration c_2, then some NaCl, x, passes through the membrane until *equilibrium* (q.v.) is established, when the *chemical potential* (q.v.) of NaCl is the same on both sides.

Initially	Na^+	zc_1	Na^+	c_2
	R^-	c_1	Cl^-	c_2
	Na^+	$zc_1 + x$	Na^+	$c_2 - x$
At equilibrium	R^-	c_1		
	Cl^-	x	Cl^-	$c_2 - x$

This condition is fulfilled when:

$$[a(Na^+)]_1\,[a(Cl^-)]_1 = [a(Na^+)]_2\,[a(Cl^-)]_2$$

assuming that in dilute solutions the activity coefficients are unity:

$$(zc_1 + x)\,x = (c_2 - x)^2$$

i.e.

$$\frac{x}{c_2} = \frac{c_2}{zc_1 + 2c_2}$$

The fraction of NaCl diffusing through the membrane (x/c_2) is the smaller, the greater the concentration of R^- and the greater its valence. If there is only a small amount of NaR present, then the distribution of NaCl, $(c_2 - x)/x$, on the two sides of the membrane is nearly 1, but if $c_1 \gg c_2$, nearly all the NaCl remains on the other side of the membrane to R^-.

At equilibrium there is an electrical potential difference, the membrane potential, caused by the establishment of a *concentration cell*†, given by

$$E = \frac{2.303\,RT}{F}\log\frac{c(\text{NaCl, RHS})}{c(\text{NaCl, LHS})} = \frac{2.303\,RT}{F}\log\frac{(c_2 - x)}{x}$$

$$= \frac{2.303\,RT}{F}\log\,(1 + zc_1/c_2)$$

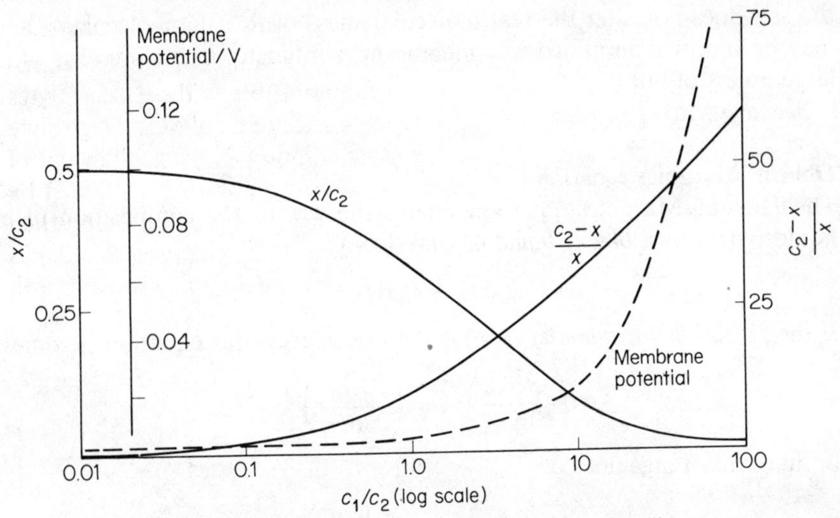

Figure D.6 Donnan membrane equilibrium curves ($z = 1$).

With a colloidal anion, E is positive, and with a colloidal cation, E is negative. With a polyvalent ion ($z > 1$), values of E are larger than in figure D.6, e.g. for $z = 10$, $c_1/c_2 = 100$, $E = 0.178$ V.

In the determination of the *osmotic pressure* (q.v.) of a colloidal electrolyte using a membrane impermeable to R^-, the unequal distribution of electrolyte may make the osmotic pressure very different from that due to the colloid alone. The observed osmotic pressure, π_{obs}, is determined by the difference in the number of particles on the two sides of the membrane; i.e., in dilute solution,

$$\pi_{obs} = RT\,\{(z+1)c_1 - 2c_2 + 4x\}$$

while the osmotic pressure, π_0, due to NaR alone is

$$\pi_0 = RT(z+1)c_1$$

Hence,

$$\frac{\pi_{obs}}{\pi_0} = 1 - \frac{2c_2 z}{(z+1)(zc_1 + 2c_2)}$$

If $c_1 \gg c_2$, $\pi_{obs} \approx \pi_0$, while if $c_2 \gg c_1$ then $\pi_{obs} = \pi_0/(z+1) = RTc_1$. Thus if the colloid is swamped with a large excess of diffusible electrolyte, π_{obs} is equal to the osmotic pressure which would have been found for an undissociated colloid ($z \approx z + 1$, when z is large). Thus, provided the presence of the salt does not alter the real molecular mass of R^-, the molecular mass may be simply determined in a membrane osmometer in the presence of a large excess of salt.

See also Mo.

Duhem–Margules equation

The Duhem–Margules equation relates the v.p. to the composition of a liquid pair. For a *binary liquid mixture* (q.v.),

$$n_A \, d\bar{J}_A + n_B \, d\bar{J}_B = 0$$

If the *partial molar quantity* (q.v.) is the free energy, this equation becomes

$$\left(\frac{\partial \bar{G}_A}{\partial \ln x_A}\right)_{T,P} = \left(\frac{\partial \bar{G}_B}{\partial \ln x_B}\right)_{T,P}$$

or, in terms of fugacities,

$$\left(\frac{\partial \ln p_A^*}{\partial \ln x_A}\right)_{T,P} = \left(\frac{\partial \ln p_B^*}{\partial \ln x_B}\right)_{T,P}$$

If the vapours behave ideally, then this equation becomes the Duhem–Margules equation:

$$\left(\frac{\partial \ln p_A}{\partial \ln x_A}\right)_{T,P} = \left(\frac{\partial \ln p_B}{\partial \ln x_B}\right)_{T,P}$$

or

$$\left(\frac{\partial p_A}{\partial x_A}\right)_{T,P} = \frac{p_A}{x_A}\frac{x_B}{p_B}\left(\frac{\partial p_B}{\partial x_B}\right)_{T,P}$$

which, if the solution is ideal, becomes

$$\left(\frac{\partial p_A}{\partial x_A}\right)_{T,P} = \frac{p_A^{\ominus}}{p_B^{\ominus}}\left(\frac{\partial p_B}{\partial x_B}\right)_{T,P}$$

This last equation shows that, for binary liquid mixtures, the curves of p_A and p_B against x_A and x_B, respectively, slope in the opposite directions.

E

Ebullioscopic constant

The ebullioscopic constant, k_e (units: K kg mol^{-1}), or molal elevation of the *boiling point* (q.v.) of a solvent A is defined by the equation

$$k_e = \frac{RT_0^2}{n_A L_e} = \frac{RT_0^2}{l_e}$$

where n_A is the number of moles of solvent in 1 kg solvent. k_e is a function of the nature of the solvent only, and insofar as the solutions are ideal, is the same for all solutes. Where a choice of solvent is possible, it is best to use that with the smallest value of l_e; in this way the largest change in temperature for a given concentration of solute is obtained.

Efficiency

The thermodynamic efficiency of a reversible Carnot *heat engine* (q.v.) is the fraction of the *heat* (q.v.) absorbed by the machine which is converted into *work* (q.v.). The efficiency is the same for all conceivable reversible engines working between the same temperature limits:

$$\text{Efficiency} = \eta = \frac{w_{max}}{q_2} = \frac{T_2 - T_1}{T_2}$$

where q_2 is the heat absorbed at the higher temperature, T_2. The efficiency is determined solely by the difference between the temperatures of the hotter and colder reservoirs and the thermodynamic temperature at which the heat is absorbed. Thus the efficiency of a steam engine operating between 120 °C and a condenser temperature of 20 °C is given by

$$\frac{393 - 293}{393} \times 100 = 25\%$$

This means that, for every 4 J of heat supplied at the higher temperature, the equivalent of 1 J of work is obtained and 3 J of heat is given off at the lower temperature. This is the maximum efficiency that could be expected if there were no inefficiencies in the operation. Additionally, there is a mechanical efficiency that limits the available work to a fraction of this theoretical amount.

Efflorescence

Efflorescence of a *salt hydrate* (q.v.) occurs when the partial water vapour pressure of the surroundings falls below the dissociation pressure of the hydrate. The hydrate then loses water and its surface becomes covered with a layer of a lower hydrate or of the anhydrous salt. Examples include $Na_2CO_3 \cdot 10H_2O$ and $Na_2SO_4 \cdot 10H_2O$.

See also Deliquescence.

Electron affinity
The electron affinity (A) of an atom is the increase in internal energy (ΔU_0) at 0 K which accompanies the addition and binding of an electron to an atom in the gaseous phase to form a negative ion in the gaseous phase, i.e.

$$X(g) + e \longrightarrow X^-(g)$$

Accurate values of electron affinities have been obtained from spectroscopic measurements. Electron affinities are usually encountered in Born–Haber (enthalpy) cycles (see *Born–Haber cycle*) which refer to constant pressure systems at 298 K. The corresponding enthalpy change (ΔH, 298 K) is related to the electron affinity by the application of *Kirchhoff's equation* (q.v.):

$$\Delta H(298\ \text{K}) = \Delta U_0 + \int_0^{298} [C_p(X^-) - C_p(X) - C_p(e)]\, dT$$

Assuming that X(g), $X^-(g)$ and e(g) can be regarded as ideal monatomic gases, then the heat capacities of these species may be taken as 0 at 0 K and $5R/2$ at other temperatures. Thus

$$\Delta H(298\ \text{K}) = \Delta U_0 - 5RT/2$$

at 298 K the correction amounts to about 6.2 kJ mol^{-1}. However, in such enthalpy cycles both an *ionisation potential* (q.v.) and an electron affinity usually contribute; if the number of electrons involved in each term is the same, then the $5RT/2$ correction terms cancel out and the ΔU_0 values can be used.

Although for the majority of elements the addition of an electron releases a large amount of energy, in several cases (Be, Mg, N, He and Ne) the addition is accompanied by an increase in energy.

See also table A.III (p. 253); and Da.

Elevation of boiling point
See Boiling point.

Ellingham diagrams
The variation of the standard free energy change (see *free energy*) of a reaction with T is almost linear. The graphical method of representing free energy data, first proposed by Ellingham, is most suitable for collating and discussing a wide range of chemical reactions. In the diagrams for oxide formation, values of $\Delta G^\ominus$, involving 1 mole of oxygen in the formation, are plotted against T. The plots are linear, and, with the one notable exception of carbon monoxide (CO), have positive slopes, defined by $(\partial(\Delta G^\ominus)/\partial T)_P = -\Delta S^\ominus$;

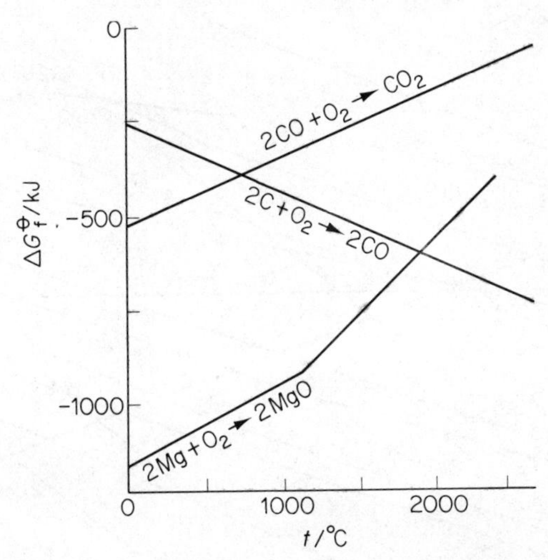

Figure E.1 Standard free energy changes per mole of O_2 for three oxidation reactions from 0 to 2500 °C.

this means that the oxides (except CO) become less stable at higher temperatures. The abrupt change of slope at the m.p. and b.p. of the element is due to the increase in entropy which occurs at each phase change. If the oxide melts or boils, this detracts from the entropy loss and the $\Delta G^{\ominus}$ plot changes the direction of slope (e.g. MnO).

By use of the simple diagram (figure E.1), $\Delta G^{\ominus}$ for a reduction can be calculated by subtracting $\Delta G^{\ominus}$ for one oxidation reaction from $\Delta G^{\ominus}$ for another at the same temperature. Thus, for the oxidations

$$2\,C + O_2 \longrightarrow CO_2 \qquad \Delta G_1^{\ominus}$$
$$2\,Mg + O_2 \longrightarrow MgO \qquad \Delta G_2^{\ominus}$$

subtraction gives

$$2C + 2MgO \longrightarrow 2CO + 2Mg \qquad \Delta G_3^{\ominus} = \Delta G_1^{\ominus} - \Delta G_2^{\ominus}$$

At low temperatures $\Delta G_3^{\ominus} > 0$, decreasing to 0 at the point of intersection of the curves; above this temperature (1900 K) $\Delta G_3^{\ominus} < 0$ and the reduction of magnesia by carbon is thermodynamically feasible. Note that MgO cannot be reduced by CO at any temperature.

By use of such diagrams, any oxidation–reduction process can be analysed visually (figure E.2); the oxide lower in the chart at a given temperature is the more stable. Any metal oxide will be reduced by a metal for which the

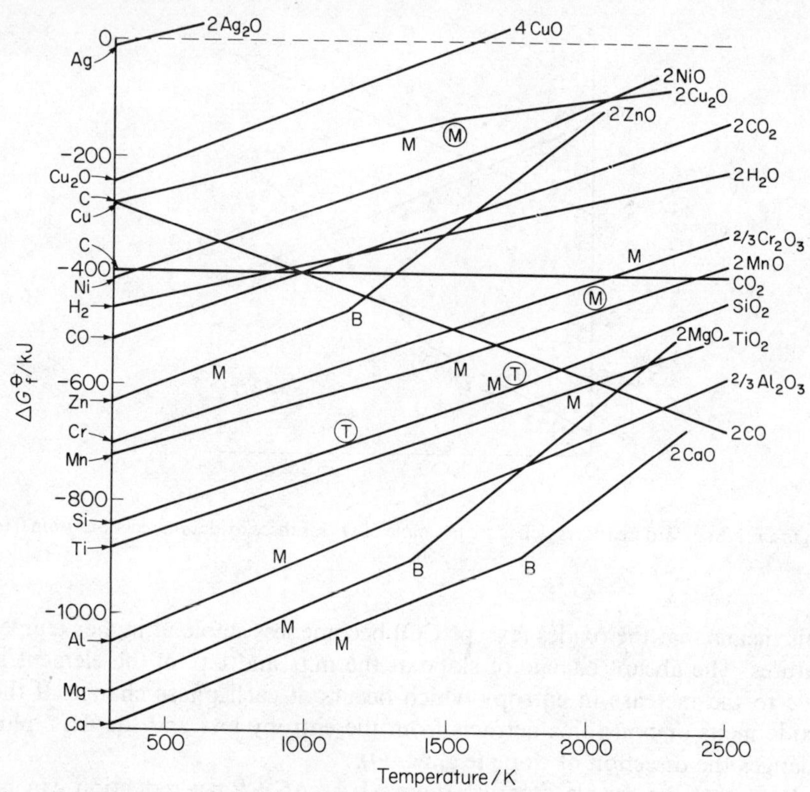

Figure E.2 Simplified Ellingham diagram for oxide formation, based on 1 mole of oxygen. M,B,T represent m.p., b.p. and transition point, respectively, for the element; a letter in a circle denotes the same point for the oxide.

$\Delta G^\ominus - T$ line for the oxide formation lies lower in the diagram (i.e. a lower metal will reduce an upper oxide).

The reducing power of H_2 is limited and does not increase markedly with temperature; it is only of use for metals above it in the chart. Carbon is clearly the most versatile reducing agent of all, and will reduce almost any oxide, provided the temperature is raised sufficiently; such practical difficulties as high cost and the stability of refractories for furnace linings limit its exploitation. Below 1000 K, CO is a more effective reducing agent than is C. When the use of C becomes impracticable, Al provides an alternative, but ultimately recourse may be had to electroreduction, where the reducing power, a function of the cathode potential, is unlimited.

The diagrams also show why gold and other precious metals exist native; their oxides can be 'reduced' merely by heating them to the temperature at which this $\Delta G_f^\ominus$ becomes zero.

For any states other than the standard states, the appropriate free energy change may be calculated from the *van't Hoff isotherm* (q.v.).

Similar Ellingham diagrams are available for sulphide and chloride formation.

See also Ellingham, H.J.T. *J. Soc. Chem. Ind.*, **63**, 125 (1944); Ives, D.J.G. (1960). *Principles of the Extraction of Metals*, R.I.C. Monograph No. 3.

Enantiotropy

Enantiotropy is exhibited by two allotropic or polymorphic forms when the transition temperature under atmospheric pressure lies below the m.p. of the solid (figure E.3). Each solid form has a definite temperature range of stability and the transition is reversible. There are four possible *triple points* (q.v.), viz. S_α S_β V, S_α L V, S_α S_β L (metastable) and S_β L V.

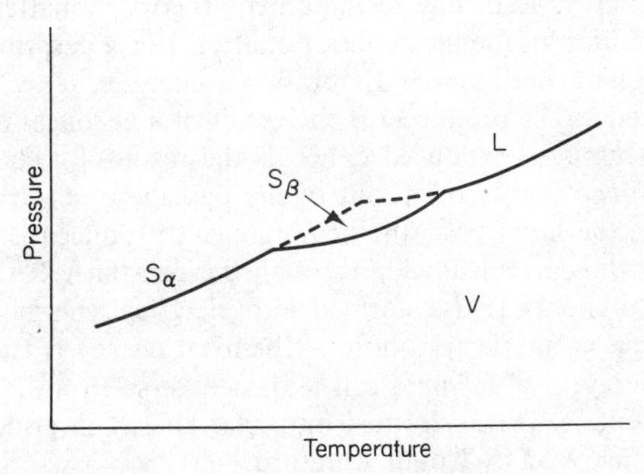

Figure E.3 Vapour pressure–temperature diagram of an enantiotropic system.

Examples of enantiotropic systems are sulphur (see *sulphur system*) tin, mercury (II) iodide and ammonium nitrate. If the polymeric substance exists in more than two crystalline forms, then it is possible for some forms to be enantiomorphic while others are monotropic.

See also Monotropy; One-component system; Phase rule; and F & C, F & J.

Endothermic process

An endothermic process, either physical or chemical, is one which is accompanied by the absorption of heat, i.e. the *enthalpy* (q.v.) of the reactants is less than that of the products (ΔH is positive). For example,

$$\tfrac{1}{2}N_2(g) + \tfrac{1}{2}O_2(g) \longrightarrow NO(g) \qquad \Delta H(2146\ K) = 90.38\ kJ$$

A reaction which is endothermic in one direction is exothermic in the reverse direction. The sign of ΔH is *no* criterion of the spontaneity of the process.

See also Exothermic process.

Energy

Energy (dimensions: m l^2 t^{-2}; units: J) exists in many different forms; however, all energy is either kinetic or potential.

Mechanical energy is the ability to do work.

Kinetic energy ($\frac{1}{2}mv^2$) depends on the motion of a body.

Potential energy (*mgh*) depends on the position of a body in a force field, e.g. a gravitational field.

Thermal energy, according to the kinetic theory of matter, is the energy of random motion of the molecules in matter. For a gas, this is the sum of the translational, rotational and vibrational energies.

Chemical energy is produced as the result of a chemical reaction.

Electrical energy is produced either as the result of a chemical reaction in a *galvanic cell*† or as the result of the passage of electricity through a resistance. In the latter case, the heating effect produced is given by I^2Rt, where I/A is the current flowing through the resistance R/Ω for t seconds.

Mechanical energy is transformed into electrical energy in a generator and vice versa in an electric motor. Chemical energy is transformed into electrical energy in a galvanic cell and vice versa in an *electrolytic cell*†. Mechanical energy is transformed into heat energy in grinding processes, as first demonstrated by Count Rumford.

Although conversion of heat into other forms of energy is limited by the *second law of thermodynamics* (q.v.), any form of energy can be completely converted into heat energy (see *heat*). Thus energy may be defined as heat or anything that can be converted into heat. All forms of energy can be considered as the product of an extensive *property* (q.v.) (e.g. S) and an intensive property (e.g. T).

The interconversion of mass and energy can be readily measured in nuclear reactions, but for normal chemical reactions, the change in mass, theoretically associated with the energy change, is so small that it cannot be measured:

$$\Delta U = c^2\ \Delta m$$

For example, for the recombination of two hydrogen atoms, $\Delta U = 432$ J mol^{-1} $= 2.16 \times 10^8$ J kg^{-1} hydrogen; for this change $\Delta m = 2.4$ μg per kg of hydrogen.

See also Heat capacity; Internal energy.

Enthalpy

Enthalpy, H (dimensions: m l^2 t^{-2}; units: J mol^{-1}), is defined by

$$H = U + PV$$

H, like U, P and V, is a *state function* (q.v.) and its value is independent of the way in which the particular state is reached. Absolute values of H are not known, since absolute values of U cannot be obtained from classical thermodynamics alone.

The heat required to increase the temperature of a substance from T to $T + \mathrm{d}T$ at constant P is

$$đq_p = \mathrm{d}H = C_p \mathrm{d}T \qquad \text{whence} \quad C_p = \left(\frac{\partial H}{\partial T}\right)_P$$

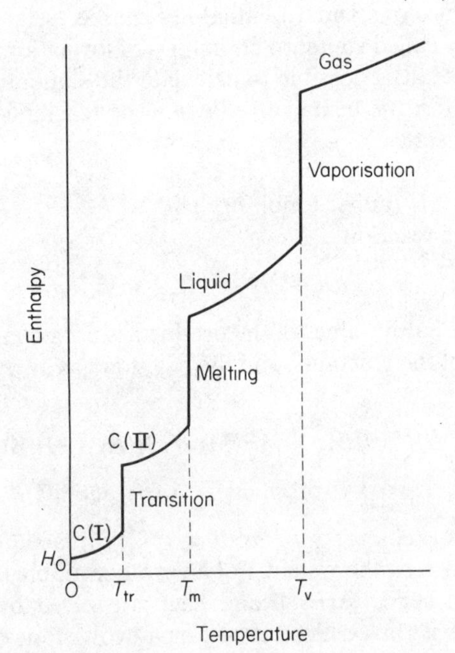

Figure E.4 Schematic diagram of enthalpy (or internal energy) of a substance as a function of temperature. (C(I) and C(II) represent crystalline forms I and II, respectively.)

Enthalpy

For an ideal gas at 0 K, $H_0 = U_0$ and thus, at T K,

$$H(T) = U_0 + \left(\frac{\partial U}{\partial T}\right)_V \mathrm{d}T + PV = U_0 + \tfrac{3}{2}RT + RT$$

For any system, $H = f(T)$, as shown in figure E.4, in which the ordinate represents H (or U) referred to its value at 0 K. Referred to the enthalpy of the crystalline form I at 0 K, the enthalpy of the substance in the gaseous form at T is given by

$$H(\mathrm{g}, T) = H_0(\mathrm{I}) + \int_0^{T_{\mathrm{tr}}} C_p(\mathrm{I})\,\mathrm{d}T + L_{\mathrm{tr}} + \int_{T_{\mathrm{tr}}}^{T_{\mathrm{m}}} C_p(\mathrm{II})\,\mathrm{d}T + L_{\mathrm{f}} + \int_{T_{\mathrm{m}}}^{T_{\mathrm{v}}} C_p(\mathrm{l})\mathrm{d}T + L_{\mathrm{e}} + \int_{T_{\mathrm{v}}}^{T} C_p(\mathrm{g})\,\mathrm{d}T$$

where T_{m}, T_{tr} and T_{v} are the melting, transition and boiling points, respectively.

Absolute enthalpy values cannot be determined, but if the standard enthalpy value of the elements in their standard states is arbitrarily taken as zero, then values of standard enthalpies can be assigned to compounds. These are usually called standard enthalpies of formation, $\Delta H_{\mathrm{f}}^{\ominus}$ (see below). From these values, it is possible to calculate the standard enthalpy change for any reaction, i.e. the heat evolved when the reactants and products are in their standard states.

Enthalpy change, ΔH (units: J mol^{-1})
In general, for the reaction

$$aA + \mathrm{b}B \longrightarrow l\mathrm{L} + m\mathrm{M}$$

the sum of the enthalpy values of the products is not equal to the sum of the enthalpy values of the reactants, and the reaction has an associated enthalpy change given by

$$\Delta H = lH(\mathrm{L}) + mH(\mathrm{M}) - aH(\mathrm{A}) - bH(\mathrm{B})$$
$$= \sum H(\text{products}) - \sum H(\text{reactants})$$

For an endothermic change, $\sum H(\text{products}) > \sum H(\text{reactants})$; hence, $\Delta H > 0$ and heat is absorbed by the system. For an exothermic change, $\sum H(\text{products}) < \sum H(\text{reactants})$; hence, $\Delta H < 0$ and heat is evolved by the system. The heat of a reaction is thus defined as the heat evolved, at constant P and T, when the amounts of material shown in the stoichiometric equation react to give products. Since ΔH depends on the physical state of the reactants and products, these must be specified. The standard enthalpy change (see below), $\Delta H^{\ominus}$, is the heat evolved when the reactants and products are in their standard states.

Although it is customary to distinguish between different types of chemical reaction, and to specify enthalpy changes of combustion, dilution, formation, neutralisation, solution and transition, all are, in fact, heats of reaction. Thus, for the reaction,

$$2H_2(g) + O_2(g) \longrightarrow 2H_2O(l) \qquad \Delta H^{\ominus}(298\ \mathrm{K}) = -571.6\ \mathrm{kJ}$$

i.e. 571.6 kJ of heat is evolved; or for the formation of liquid water at 298 K, $\Delta H_f^{\ominus} = -285.8\ \mathrm{kJ\ mol^{-1}}$; or for the combustion of hydrogen at 298 K, $\Delta H_c^{\ominus} = -285.8\ \mathrm{kJ\ mol^{-1}}$.

For a series of chemical reactions ΔH values are additive (see *Hess's law of constant heat summation*). ΔH for a chemical reaction varies with temperature according to *Kirchhoff's equation* (q.v.):

$$\Delta H(T_2) = \Delta H(T_1) + \Delta C_p\, \mathrm{d}T$$

ΔH is related to other thermodynamic quantities by the following equations:

$$\Delta H = \Delta G + T\Delta S; \qquad \mathrm{d}H = V\,\mathrm{d}P + T\,\mathrm{d}S$$

$$\left(\frac{\partial H}{\partial P}\right)_S = V; \qquad \left(\frac{\partial H}{\partial S}\right)_P = T$$

$$\frac{\Delta H^{\ominus}}{RT^2} = \left(\frac{\partial \ln K_p}{\partial T}\right)_P$$

$$\Delta H + w' = T\left(\frac{\partial w'}{\partial T}\right)_P \quad \text{or} \quad \Delta H - \Delta G = -T\left(\frac{\partial(\Delta G)}{\partial T}\right)_P$$

$$\Delta H + nFE = nFT\left(\frac{\partial E}{\partial T}\right)_P$$

Determination of ΔH *values.* ΔH values can be determined in the following ways:

(1) Calorimetric method for reactions in solution: the change in T is measured when the reactants are mixed in an insulated container. From a knowledge of the heat capacity of the calorimeter, the heat of reaction can be calculated (see *thermometric titrations*).

(2) The calorimetric method for spontaneous and rapid reactions in which

at least one reactant is gaseous makes use of the *bomb calorimeter* (q.v.). The heat evolved is measured by the change in T of the surrounding water and the bomb. This method gives the heat content change at constant V, i.e. ΔU, from which ΔH can be calculated, since $\Delta H = \Delta U + \Delta nRT$. The method is suitable for heats of reaction, combustion and formation.

(3) The electrochemical method for ionic reactions in solution makes use of a *reversible cell*†, so constructed that the required reaction is the cell reaction. The variation of e.m.f. with T is measured and ΔH calculated from the modified *Gibbs–Helmholtz equation* (q.v.):

$$\Delta H = -nFE + nFT\left(\frac{\partial E}{\partial T}\right)_P$$

The heat of reaction can usually be determined more accurately from the e.m.f. and its temperature coefficient than by direct calorimetry, provided all necessary precautions are taken.

(4) Indirectly, using Hess's law.

(5) From a knowledge of ΔG and ΔS for the reaction

$$\Delta H = \Delta G + T\Delta S$$

Standard enthalpy change

The standard enthalpy change, $\Delta H^{\ominus}$, is the enthalpy change when the reactants and products are in their standard states.

Determination of $\Delta H^{\ominus}$ values $\Delta H^{\ominus}$ values can be determined in the following ways:

(1) From the *van't Hoff isochore* (q.v.) and the measured variation of K with temperature. It is suitable for reactions involving gases.

(2) From a knowledge of $\Delta G^{\ominus}$ and $\Delta S^{\ominus}$ for the reaction.

(3) From tabulated values of standard heats of formation or combustion of reactants at a specified temperature:

$$\Delta H^{\ominus} = \sum \Delta H_f^{\ominus}(\text{products}) \quad - \sum \Delta H_f^{\ominus}(\text{reactants})$$

$$\Delta H^{\ominus} = \sum \Delta H_c^{\ominus}(\text{reactants}) \quad - \sum \Delta H_c^{\ominus}(\text{products})$$

Values of $\Delta H_f^{\ominus}$ are generally listed at one temperature only; for values at other temperatures, it is necessary to use *heat capacity* (q.v.) equations of the type

$$\Delta H^{\ominus}(T\ \text{K}) = \Delta H_0^{\ominus} + \Delta aT + \frac{\Delta b}{2}T^2 + \frac{\Delta c}{3}T^3 + \cdots$$

(4) With the increasing use of statistical thermodynamics, another method of tabulation, using $H_T^\ominus - H_0^\ominus$, or the *enthalpy function* (q.v.), $(H_T^\ominus - H_0^\ominus)/T$, has come into general use for ideal gases. This method of tabulation avoids the use of empirical heat capacity equations and makes for more ready comparison of data from different sources. Thus the heat of formation of a compound from its elements can be calculated by using the equation

$$\Delta H_f^\ominus(T\ \mathrm{K}) = \Delta H_f^\ominus(0\ \mathrm{K}) + (H_T^\ominus - H_0^\ominus)_{\mathrm{compound}} - \sum(H_T^\ominus - H_0^\ominus)_{\mathrm{elements}}$$

The values of all the quantities on the right-hand side are tabulated in table A.V (p. 254).

Standard enthalpy of combustion

The standard enthalpy of combustion, $\Delta H_c^\ominus$ (units: kJ mol^{-1}), is the heat evolved when 1 mole of the substance, in its standard state, is completely burnt in the presence of excess oxygen at constant P and T.

Values of the enthalpy of combustion are calculated from measurements made with a bomb calorimeter. Standard heats of reaction can be calculated from standard enthalpies of combustion of reactants and products:

$$\Delta H^\ominus = \sum \Delta H_c^\ominus(\mathrm{reactants}) - \sum \Delta H_c^\ominus(\mathrm{products})$$

e.g., for the reaction

$$\mathrm{C(graphite)} + \tfrac{1}{2}\mathrm{O_2(g)} \longrightarrow \mathrm{CO(g)}$$

$$\Delta H^\ominus = \Delta H_c^\ominus(\mathrm{graphite}) - \Delta H_c^\ominus(\mathrm{CO}) = -393.5 - (-283.0) = -110.5\ \mathrm{kJ}$$

Standard enthalpy of formation

The standard enthalpy of formation, $\Delta H_f^\ominus$ (units: kJ mol^{-1}), is the heat evolved when 1 mole of the substance is formed from its elements in their standard states at a specified temperature; e.g., for the reaction

$$\underset{\text{(graphite, 298.15 K, 1 atm)}}{\mathrm{C}} + \underset{\text{(g, 298.15 K, 1 atm)}}{\mathrm{O_2}} \longrightarrow \underset{\text{(g, 298.15 K, 1 atm)}}{\mathrm{CO_2}}$$

$\Delta H^\ominus = -393.5$ kJ. When 1 mole (0.012 kg) of graphite is completely burnt in excess oxygen at 1 atm pressure and at 298.15 K to give 1 mole (0.044 kg) of gaseous carbon dioxide, 393.5 kJ of heat is evolved. Since the elements C and O_2 are in their standard states, their standard enthalpies are arbitrarily taken as zero; hence,

$$\Delta H^{\ominus} = \Delta H_f^{\ominus}(CO_2) - \Delta H_f^{\ominus}(C) - \Delta H_f^{\ominus}(O_2)$$

whence $\Delta H_f^{\ominus}(CO_2) = -393.5$ kJ at 298.15 K.

The enthalpy of formation of the members of the homologous series of n-paraffins (C_nH_{2n+2}), the olefines (C_nH_{2n}) and the primary aliphatic alcohols ($C_nH_{2n+1}OH$) increases regularly; for all compounds where $n > 5$, the increase in enthalpy for each additional CH_2 is 20.61 kJ regardless of the terminal group:

$$\Delta H_f^{\ominus} = A - 20.61\, n \text{ at } 298 \text{ K}$$

A (the intercept) is -43.56, 81.96 and -205.9 kJ mol^{-1} for the paraffins, olefines and alcohols, respectively.

Standard heats of reaction can be calculated from tabulated values (table A.I, p. 247) of standard enthalpy

$$\Delta H^{\ominus} = \sum \Delta H_f^{\ominus}(\text{products}) - \sum \Delta H_f^{\ominus}(\text{reactants})$$

Standard enthalpy of an ion

The standard enthalpy of an ion, $\Delta H_f^{\ominus}$, is a concept which only applies to ionic reactions in infinitely dilute solution and provides a means of tabulating standard enthalpies of formation of individual ions.

The solution of HCl(g) in water can be represented (all data at 298.15 K) thus:

$$HCl(g) + aq \longrightarrow H_3O^+(aq) + Cl^-(aq) \qquad \Delta H^{\ominus} = -75.1 \text{ kJ}$$

But $\Delta H_f^{\ominus}(HCl, g) = -92.30$ kJ; thus

$$\Delta H^{\ominus} = -75.1 = \Delta H^{\ominus}(H_3O^+, aq + Cl^-, aq) - \Delta H_f^{\ominus}(HCl, g)$$

or

$$\Delta H_f^{\ominus}(H_3O^+, aq + Cl^-, aq) = -75.1 - 92.3 = -167.4 \text{ kJ mol}^{-1}$$

Since no reaction will produce a single ionic species (solutions are always electrically neutral), it appears that $\Delta H_f^{\ominus}$ can only be listed for pairs of ions. The enthalpy of a dilute solution can, however, be treated in terms of the enthalpy contributions of the separate ionic species. Thus, if an arbitrary value is attributed to one ion, then values can be assigned to $\Delta H_f^{\ominus}$ for all other ions. The accepted reference is the proton, for which $\Delta H_f^{\ominus}(H_3O^+, aq)$ is assigned the value 0. Thus $\Delta H_f^{\ominus}(Cl^-, aq) = -167.4$ kJ mol^{-1}, and the values for all other ions follow immediately.

For the reaction

$$Ba(c) + 2H^+(aq) \longrightarrow H_2(g) + Ba^{2+}(aq) \quad \Delta H^{\ominus} = -538.4 \text{ kJ}$$

it follows that $\Delta H^\ominus = \Delta H_f^\ominus(Ba^{2+}, aq) = -538.4$ kJ mol^{-1}, since Ba and H_2 are elements with zero enthalpies of formation and $H_3O^+(aq)$ is the standard reference.

From such tabulated data (table A.I, p. 247) it is possible to calculate enthalpy changes in ionic reactions in solution, e.g. the neutralisation of 1 mole of an acid with 1 mole of a base, which in dilute solution can be represented:

$$H_3O^+(aq) + OH^-(aq) \longrightarrow H_2O$$

for which

$$\begin{aligned}\Delta H^\ominus &= \Delta H_f^\ominus(H_2O, l) - \Delta H_f^\ominus(H_3O^+, aq) - \Delta H_f^\ominus(OH^-, aq)\\ &= -285.8 - 0 - (-229.9) = -55.9 \text{ kJ}\end{aligned}$$

a value independent of the nature of the acid or base. Values of heats of neutralisation less than this are obtained when weak acids or bases are involved, owing to incomplete dissociation of the acid or base; some of the heat energy is required to complete the dissociation.

Standard enthalpy of solution

The standard enthalpy of solution of any substance is the heat evolved at a specified T when 1 mole of the substance is dissolved in such a large amount of solvent that further dilution causes no measurable thermal effect.

Mixing of two liquids which form an ideal solution causes no change in H; the mixing of many pairs of organic liquids fulfils this condition. Solutions in water and other polar solvents involve enthalpy changes, especially when the solute is an electrolyte. Behaviour is complicated by interionic effects and by possible interaction between the solute and solvent, e.g. solvation. Values of ΔH for the solution of salts in water vary greatly, but, in general, fully hydrated salts and those which form no crystalline hydrates dissolve endothermally, while anhydrous salts that can form hydrates dissolve exothermally. When 1 mole of HCl gas is dissolved in a very large excess of water, 75.1 kJ is evolved; this is expressed:

$$HCl(g) + aq \longrightarrow HCl(aq) \qquad \Delta H^\ominus(298\text{ K}) = -75.1 \text{ kJ}$$

The heat evolved varies with the dilution, but the over-all heat of solution is equal to the algebraic sum of the separate heats of solution and dilution:

$$CaCl_2(c) + 400\,H_2O \longrightarrow CaCl_2{\cdot}400\,H_2O(l) \qquad \Delta H = -75.69 \text{ kJ} \quad (E.1)$$
$$CaCl_2{\cdot}400\,H_2O(l) + aq \longrightarrow CaCl_2(aq) \qquad \Delta H = -\ \ 1.8 \text{ kJ} \quad (E.2)$$
$$CaCl_2 + aq \longrightarrow CaCl_2(aq) \qquad \Delta H = -77.49 \text{ kJ} \quad (E.3)$$

Equation (E.1) represents the solution of 1 mole of $CaCl_2$ in 400 moles of water ($CaCl_2 \cdot 400\ H_2O$ is not a defined chemical compound, but defines the amount of water in which the solid is dissolved). Equation (E.2) shows the heat evolved when this solution is further diluted in a very large amount of water, while equation (E.3), the sum of the previous equations, gives the over-all heat of solution of 1 mole of $CaCl_2$ in a very large quantity of water.

Heats of solution and dilution are measured in an adiabatic calorimeter.

Heat (molar) of transition

The heat (molar) of transition, or latent heat, is the heat evolved when 1 mole of the substance changes isothermally, at a specified T, from the form (I) stable at the lower T to the form (II) stable at the higher T. Heat is always absorbed in these transformations, i.e. $\Delta H_{tr} > 0$.

The normal heats of transition are those for fusion (L_f) at the m.p., vaporisation (L_e) at the b.p., sublimation (L_s) and transition between two allotropic forms (L_{tr}) at the transition temperature.

Heats of transition, like all values of ΔH, vary with T; the value at any other T can be calculated by using *Kirchhoff's equation* (q.v.) and known values of ΔC_p (i.e. $C_p(\text{II}) - C_p(\text{I})$).

Determination of heats of transition Heats of transition can be determined in the following ways:

(1) L_e and L_s can be obtained from a study of the variation of the v.p. of a liquid or a solid, respectively, with T and the application of the *Clausius–Clapeyron equation* (q.v.). The slope of the log P–T^{-1} curve at a given T is $-L_e/2.303\ R$.

(2) L_{tr} can be calculated from $\Delta H_f^\ominus$ or $\Delta H_c^\ominus$ values; e.g., at 298 K,

$$\begin{array}{lll} \text{C(graphite)} + O_2(g) \longrightarrow CO_2(g) & & \Delta H = -393.513\ \text{kJ} \\ \text{C(diamond)} + O_2(g) \longrightarrow CO_2(g) & & \Delta H = -395.409\ \text{kJ} \\ \qquad \text{C(diamond)} \quad \longrightarrow \text{C(graphite)} & & \Delta H = 1.896\ \text{kJ} \end{array}$$

See also table A.I (p. 247); and Bar, Da, Dic, G & S, K, Ro, Wal, War.

Enthalpy function

The enthalpy function (units: J K^{-1} mol^{-1}) for an ideal gas is given by

$$\frac{H_T^\ominus - H_0^\ominus}{T} = \frac{5}{2}R + RT\left(\frac{\partial \ln Q_i}{\partial T}\right)_P$$

Tabulated values (table A.V, p. 254), which are obtained from spectroscopic measurements, are of great use in the calculation of standard enthalpy changes at any temperature. The advantage of this method of calculating $\Delta H_f^\ominus$ is that it avoids the use of empirical heat capacity data.

See also Partition function; and K.

Entropy

Entropy, S (dimensions: m l^2 t^{-2} deg^{-1}; units J K^{-1} mol^{-1}), an extensive property of a system, was first introduced by Clausius, who defined the increase in entropy of an isothermal reversible process as

$$\Delta S = q_{rev}/T$$

or, for an infinitesimal reversible process,

$$dS = đq_{rev}/T$$

and, for a non-isothermal reversible process,

$$\Delta S = \sum dS = \int đq_{rev}/T$$

where q_{rev} is the heat absorbed when the process is conducted reversibly and depends only on the initial and final states of the system. If the process is conducted irreversibly, then $q/T < \Delta S$, since $q < q_{rev}$.

The entropy change in a reversible *Carnot cycle* (q.v.) is zero, i.e.

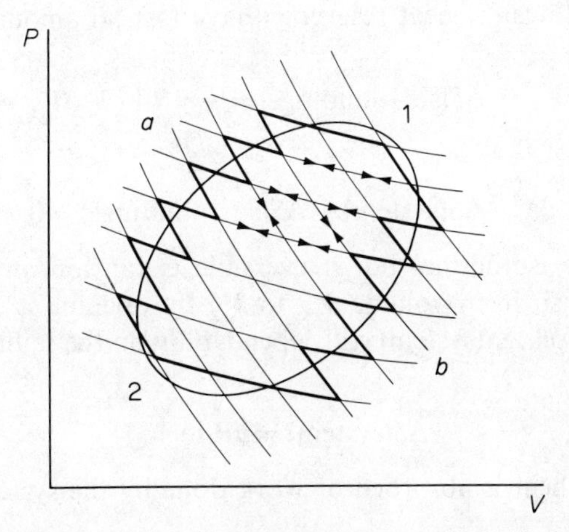

Figure E.5 Cyclic process 1–2–1 resolved into a large number of simple Carnot cycles, each having two isothermal and two adiabatic changes. The larger the number of Carnot cycles the nearer will the outside boundaries of these approximate to the closed cycle 1–2–1.

$$\oint đq_{rev}/T = 0$$

Any complete cycle can be reduced to a very large number of infinitesimal Carnot cycles (figure E.5), and consequently the entropy change in any complete cycle is zero, i.e.

$$\Delta S_a(1 \rightarrow 2) + \Delta S_b(2 \rightarrow 1) = 0$$

Thus the entropy at state 2 is independent of the path by which state 2 has been attained; hence, S, like U, H and G, is a *state function* (q.v.). For the process 1 to 2,

$$\Delta S = S_2 - S_1 = \int_1^2 đq_{rev}/T$$

This equation can be used to calculate ΔS.

The system and surroundings must be clearly distinguished, as also must the reversible or irreversible nature of the process, e.g. in the isothermal reversible expansion of n mole of a perfect gas from volume V_1 to V_2 for which $\Delta U = 0$,

$$đq = đw = nRT \ln V_2/V_1$$

Hence,

$$\Delta S(\text{system}) = đq/T = nR \ln V_2/V_1, \text{ i.e. } \Delta S(\text{system}) > 0$$

The surroundings (or heat reservoir) have lost an amount of heat $đq$ and, hence,

$$\Delta S(\text{surroundings}) = -nR \ln V_2/V_1$$

whence

$$\Delta S(\text{system}) + \Delta S(\text{surroundings}) = 0$$

For the same isothermal but irreversible expansion, in which the gas is allowed to rush from volume V_1 to V_2 by opening a stop-cock into an evacuated vessel, ΔS(system) still dependsonly on the initial and final states; hence,

$$\Delta S(\text{system}) = nR \ln V_2/V_1$$

but, since no heat is absorbed or work done by the system,

$$\Delta S(\text{surroundings}) = 0$$

i.e. $\Delta S(\text{system}) + \Delta S(\text{surroundings}) = nR \ln V_2/V_1 > 0$

In all irreversible processes there is an over-all increase in entropy; since all natural processes are spontaneous, they must occur with an increase of entropy and therefore the total entropy of the universe is continually increasing (Clausius' statement of the *second law of thermodynamics*, q.v.).

The net effect of a *spontaneous process* (q.v.), e.g. the diffusion of a solute from a higher to a lower concentration, is a running-down process making some of the energy of the universe unavailable. Thus, although the energy of the universe is constant (*first law of thermodynamics*, q.v.), it is running down and the energy is becoming less available for useful purposes (according to the second law). In any change, ΔS(total) gives a quantitative measure of the degradation of energy, and, for a naturally occurring process, ΔS(total) > 0. For a reaction (or process) and its surroundings, the following generalisations apply:

(1) $\Delta S(\text{total}) > 0$; reaction proceeds spontaneously from state 1 to 2
(2) $\Delta S(\text{total}) = 0$; no spontaneous reaction will occur
(3) $\Delta S(\text{total}) < 0$; reaction proceeds spontaneously in the reverse direction, i.e. state 2 to 1

where

$$\Delta S(\text{total}) = S_2(\text{total}) - S_1(\text{total})$$

Consider the system, near the melting point,

$$H_2O(s) \rightleftharpoons H_2O(l) \qquad \Delta H = 6026\ \text{J}$$

S/J K^{-1}	T/K	263	273	283
$\Delta S(\text{system}) = S(\text{l}) - S(\text{s})$		22.07	22.07	22.07
$S(\text{surr.}) = -\Delta H/T$		−22.91	−22.07	−21.29
ΔS(total)		− 0.84	0	+ 0.76
		Melting impossible	Melting just possible	Melting spontaneous

At temperatures above (below) 273 K the molar entropy values of the liquid and solid forms are greater (less) than the value at 273 K, but the difference between them can be assumed to be constant. Similarly, over such a small temperature range, ΔH can be assumed to be constant; the entropy decrease of the surroundings ($-\Delta H/T$) will vary with temperature, as shown in the second line of the table, and, hence, ΔS(total) varies (third line). Broadly speaking, cold water below 273 K freezes and ice above 273 K melts because both processes result in an increase in the entropy of the universe.

To use this criterion of spontaneity of a process, it is necessary to be able to evaluate all entropy changes, i.e. of both system and surroundings; the latter is not always known. A better test is the *free energy function* (q.v.), G.

S or ΔS is related to other thermodynamic quantities by the following equations:

$$G = H - TS \qquad \text{or} \qquad \Delta G = \Delta H - T\Delta S$$

$$\Delta S^{\ominus} = -\left(\frac{\partial(\Delta G^{\ominus})}{\partial T}\right)_P = nF\left(\frac{\partial E^{\ominus}}{\partial T}\right)_P$$

$$A = U - TS$$

$$S = -\left(\frac{\partial G}{\partial T}\right)_P \quad ; S = -\left(\frac{\partial A}{\partial T}\right)_V$$

$$S = N_A k \ln Q + N_A kT\left(\frac{\partial \ln Q}{\partial T}\right)_P$$

Determination and calculation of entropy values and entropy changes

Physical changes (1) In isothermal changes, e.g. at transition temperatures, $\Delta S\,(=\Delta H/T)$ can be calculated from a knowledge of the isothermal enthalpy change and the temperature at which the change occurs.

(2) In the expansion of n mole of a perfect gas from $P_1V_1T_1$ to $P_2V_2T_2$, the increase in entropy is given by

$$\Delta S = nC_V \ln \frac{T_2}{T_1} + nR \ln \frac{V_2}{V_1} = nC_V \ln \frac{T_2}{T_1} - nR \ln \frac{P_2}{P_1}$$

assuming C_V is independent of temperature. At constant pressure, the entropy change accompanying a change of temperature becomes

$$\Delta S = nC_V \ln T_2/T_1$$

and for an expansion at constant temperature,

$$\Delta S = nR \ln V_2/V_1 = -nR \ln P_2/P_1$$

(3) For any non-isothermal process, e.g. at constant pressure,

$$\Delta S = S(T_2) - S(T_1) = \int_{T_1}^{T_2} đq/T = \int_{T_1}^{T_2} C_p \,\mathrm{d} \ln T$$

Hence,

$$S(T_2) = S(T_1) + \int_{T_1}^{T_2} C_p \, d \ln T$$

A similar calculation can be made at constant volume, involving C_V. This method, applicable to a solid, liquid or gas, requires a knowledge of the

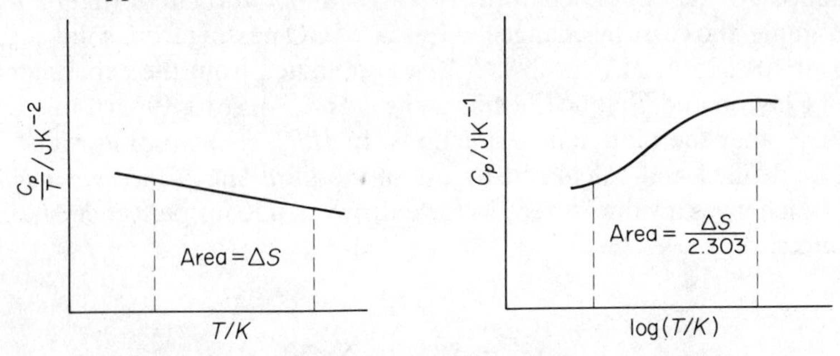

Figure E.6 Calculation of entropy change caused by increase of temperature.

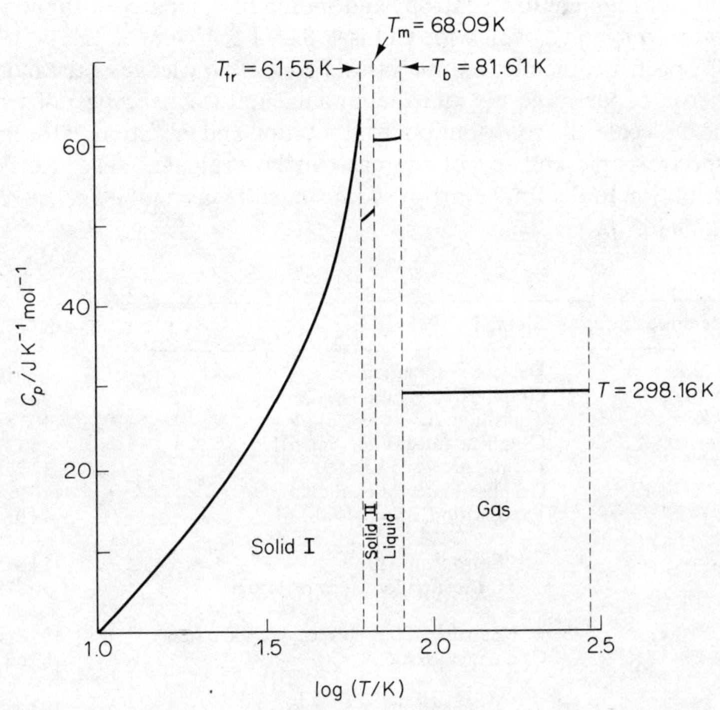

Figure E.7 C_p–log T plot for carbon monoxide.

variation of C_p with T over the range of temperatures required; the integration is usually carried out graphically (figure E.6).

If there are any isothermal transition points (e.g. m.p., b.p.) between T_1 and T_2, then the isothermal ΔS for each of these (L_{tr}/T) must be evaluated and added to that obtained from the area under the curve (figure E.7). For example, the entropy change for 1 mole of CO passing from solid at 0 K to gas at 101 325 N m^{-2} at 298.15 K was obtained from the experimental data of Clayton and Giaque (*J. Amer. Chem. Soc.*, **54**, 2610, 1932) (table E.1).

Unlike other thermodynamic functions (U, H, G, etc.), absolute values of S can be defined and calculated by use of the *third law of thermodynamics* (q.v.), which gives a value of zero to the entropy at 0 K for perfect crystalline substances.

Thus

$$\Delta S^{\ominus} = S^{\ominus}(T) - S_0$$

Hence, $S^{\ominus}$(CO, g, 298.15 K) = 193.36 J K^{-1} mol^{-1}; this is known as the thermal or *third law* (q.v.) entropy and should be compared with the *spectroscopic entropy* (q.v.), which for CO is 198.28 J K^{-1} mol^{-1}.

(4) Spectroscopic method for gases. From a knowledge of the moment(s) of inertia of the molecule and the fundamental frequency(ies) of vibration of the molecule, the contribution from rotation and vibration to the absolute or spectroscopic entropy of the gas can be evaluated. The translational contribution to the total entropy is calculated by using the *Sackur–Tetrode equation* (q.v.).

Table E.1

Temperature range	Method	ΔS/J K^{-1} mol^{-1}
0–11.70 K	Debye extrapolation	1.916
11.70–61.55 K	Graphical integration (Solid I)	40.30
61.55 K	Transition, $\Delta S = 633.0/61.55$	10.28
61.55–68.09 K	Graphical integration (Solid II)	5.139
68.09 K	Fusion, $\Delta S = 835.8/68.09$	12.27
68.09–81.61 K	Graphical integration (liquid)	10.92
81.61 K	Vaporisation, $\Delta S = 6040/81.61$	74.01
	Entropy at b.p.	154.835
	Correction for gas imperfection	0.879
	Entropy at b.p. corrected to ideal state	155.714
81.61–298.15 K	Calculated from C_p	37.65
	Total entropy of CO gas	193.364

Chemical changes (1) $\Delta S^\ominus$ for a chemical reaction can be calculated from the absolute values (either third law or spectroscopic) of the entropy of reactants and products:

$$\Delta S^\ominus = \sum S^\ominus(\text{products}) - \sum S^\ominus(\text{reactants})$$

(2) From a knowledge of the *free energy* (q.v.) change and the *enthalpy* (q.v.) change,

$$\Delta S = (\Delta H - \Delta G)/T \quad \text{or} \quad \Delta S^\ominus = (\Delta H^\ominus - \Delta G^\ominus)/T$$

(3) From a knowledge of the variation of the equilibrium constant of the reaction with the temperature or the variation of the e.m.f. of the appropriate reversible *galvanic cell*† with temperature,

$$\Delta S^\ominus = -\left(\frac{\partial(\Delta G^\ominus)}{\partial T}\right)_P = R\left(\frac{\partial(T \ln K)}{\partial T}\right)_P = nF\left(\frac{\partial E^\ominus}{\partial T}\right)_P$$

This gives the entropy change per mole of reaction.

Molar entropy values
Hard abrasive substances such as diamond, silicon carbide and quartz, in which the individual atoms are bound to one another in nearly infinite three-dimensional lattices by chemical bonds which limit the random motion of the atoms, have small entropy values. On the other hand, soft substances, such as liquids and gases, having large amounts of thermal disorder at room temperature, have large measured entropy values (diamond 2.44, platinum 41.84, lead 64.0, water 69.9, nitrous oxide 220 J K^{-1} mol^{-1}). Complex substances have larger entropy values than simple substances of similar hardness (copper 33.3, sodium chloride 72.4, zinc chloride 109 J K^{-1} mol^{-1}). Melting and vaporisation are always accompanied by an increase in entropy, e.g. the entropies of ice, water and water vapour at 273 K and 1 atm pressure are 41, 63.2 and 186 (extrapolated value) J K^{-1} mol^{-1}.

The form of the substance which exists at the higher temperature always has the highest molar entropy. The intermediate entropy of the liquid phase is always numerically closer to the entropy of the solid than to the larger entropy of the vapour (see values for water above). The inference is that liquids are more solid-like than gas-like; this is supported by the familiar view that individual atoms and molecules in solids and liquids are constantly in contact with their nearest neighbours. In contrast, the molecules in a gas have much greater freedom. The total entropy of a gas can be divided into three parts, viz. the translational, rotational and vibrational entropy.

Entropy and probability
All the evidence supports the fact that an increase in entropy is accompanied by an increase (decrease) in the disorder (order) of the system, e.g. at a m.p. or b.p., or during the isothermal expansion of a gas. This qualitative concept is often of use in assessing the entropy change in a given process or reaction, e.g.

$N_2(g) + 3H_2(g) \longrightarrow 2NH_3(g)$	For the forward reaction $\Delta S < 0$, there is a decrease in entropy since there is an increase in the order of the system.
$H_2(g) + Cl_2(g) \longrightarrow 2HCl\ (g)$	ΔS is low, since there is no change in the total number of molecules; there is a change in the type of molecule.
$2KClO_3(s) \longrightarrow 2KCl(s) + 3O_2(g)$	$\Delta S > 0$, since there has been an increase in the total number of molecules as a result of reaction.

The *Boltzmann equation* (q.v.):

$$S = k \ln W$$

provides the relationship between the entropy and W the probability or number of ways of arranging the molecules among the different energy levels. This equation provides the link between *statistical mechanics* (q.v.) and thermodynamics, and makes possible the calculation of all the thermodynamic functions of a system in terms of the properties of the microsystems of which it is composed.

As an example, the entropy change accompanying the isothermal expansion of a gas can be simply obtained on probability grounds. Consider N gas molecules in a bulb connected to an evacuated bulb of equal size but separated from it by a stop-cock. When the tap is opened, the molecules distribute themselves equally between the two bulbs. The probability that any given molecule is in one bulb is $\frac{1}{2}$ and the probability that all are in one bulb is $(\frac{1}{2})^N$. As N becomes very large, the probability of finding this situation obtaining is vanishingly small. Thus, from the molecular point of view, the gas spontaneously distributes itself uniformly, since in so doing it passes from an unlikely arrangement (all in one bulb) to a very likely one (evenly distributed).

Applying this to the isothermal expansion of 1 mole (N_A molecules) of an ideal gas from volume V_1 to volume V_2:

Probability of finding N_A molecules in final volume $V_2 = (1)^{N_A}$

Probability of finding N_A molecules in initial volume $V_1 = (V_1/V_2)^{N_A}$

Hence, the entropy finally and initially is given by

$$S_2 = k \ln (1)^{N_A} \qquad \text{and} \quad S_1 = k \ln (V_1/V_2)^{N_A}, \text{ respectively}$$

whence,

$$\Delta S = S_2 - S_1 = kN_A \{\ln 1 - \ln (V_1/V_2)\} = R \ln (V_2/V_1)$$

in agreement with the value obtained from conventional thermodynamic considerations.

Entropy of fusion, vaporisation and transition (units: J K^{-1} mol^{-1})
Fusion, vaporisation and transition are isothermal changes for which $\Delta S = \Delta H/T$, where ΔH is the appropriate heat of transition at the transition temperature T. Thus, for the melting of 1 mole of ice at 273.16 K,

$$\Delta S = \frac{L_f}{T} = \frac{6026}{273.16} = 22.05 \text{ J K}^{-1} \text{ mol}^{-1}$$

This value is the difference between the entropy of water and ice at 0 °C, i.e.

$$S(\text{l}) - S(\text{c}) = 22.05 \text{ J K}^{-1} \text{ mol}^{-1}$$

Similarly, at the boiling point, $\Delta S = L_e/T$ (*Trouton's rule*, q.v.).

Melting and vaporisation are always accompanied by an increase in the molar entropy due to an increase in the disorder of the system which accompanies the change. For non-hydrogen-bonded substances,

$$\Delta S(\text{melting}) \approx 8\text{–}12 \text{ J K}^{-1} \text{ mol}^{-1} \text{ and } \Delta S(\text{vaporisation}) \approx 90 \text{ J K}^{-1} \text{ mol}^{-1};$$

for hydrogen-bonded molecules (e.g. water), these values are somewhat higher, since the molecules in the solid (liquid) are more ordered than normally.

Standard entropy of an ion
The standard entropy of an ion, $\bar{S}_i^\ominus$, is the partial molar entropy of an ion in solution, relative to the conventional chosen standard entropy of the hydrogen ion taken as zero, i.e. $\bar{S}^\ominus(H_3O^+, aq) = 0$. It is thus possible to obtain $\bar{S}_i^\ominus$ values for other ions. Thus, for the chloride ion: for the formation of aqueous H^+ and Cl^- according to the equation

$$\tfrac{1}{2}H_2(g) + \tfrac{1}{2}Cl_2(g) \longrightarrow HCl(aq)$$

$\Delta G^\ominus = -131.265$ kJ and $\Delta H^\ominus = -167.109$ kJ; hence $\Delta S^\ominus = -120.24$ J K^{-1}

at 298 K. At this temperature, $S^\ominus(H_2, g)$ and $S^\ominus(Cl_2, g)$ from third law and spectroscopic measurements are 130.67 and 223.007 J K^{-1} mol^{-1}, respectively; hence for the above reaction,

$$\Delta S^\ominus = \bar{S}^\ominus(H_3O^+, aq) + \bar{S}^\ominus(Cl^-, aq) - \tfrac{1}{2}S^\ominus(H_2, g) - \tfrac{1}{2}S^\ominus(Cl_2, g) = -120.24$$

whence $\bar{S}^\ominus(Cl^-, aq) = 56.60$ J K^{-1} mol^{-1}.

$\bar{S}^\ominus(Zn^{2+}, aq)$ may be readily computed from the variation of $E^\ominus(Zn^{2+}, Zn)$ with temperature and the known values of $S^\ominus(Zn, s)$ and $S^\ominus(H_2, g)$ by considering the reaction

$$Zn(s) + 2H^+(aq) \longrightarrow Zn^{2+}(aq) + H_2(g)$$

For a given charge, $\bar{S}_i^\ominus$ increases with relative atomic mass, and for an approximately constant relative atomic mass, e.g. Na^+, Mg^{2+} and Al^{3+}, $\bar{S}_i^\ominus$ decreases with increase in charge.

Ionic entropy values provide a measure of the ordering effect produced by an ion on the surrounding water molecules. Small ions, e.g. Li^+, F^-, have lower entropy values than larger ions, e.g. Na^+, Cl^-. Multiply charged ions have especially low values as a result of the strong electrostatic attraction for water dipoles, and their correspondingly large ordering effect on water.

From tables of $\bar{S}_i^\ominus$ (table A.I, p. 247), it is possible to estimate entropy changes which cannot be obtained at all, or not with any reasonable precision, by direct experiment.

Entropy of mixing

A mixture of two components (either ideal gases or liquids) corresponds to a state of higher probability than do the separate components. If n_A molecules of A and n_B molecules of B are mixed to give 1 mole of a two-component system (i.e. $n_A + n_B = N_A$), then, in the mixture,

$$W = N_A!/n_A!\,n_B!$$

whereas $W = 1$ for the individual pure components, since the molecules of A (B) are indistinguishable from one another. Hence,

$$\begin{aligned}\Delta S(\text{mixing}) &= k \ln (N_A!/n_A!\,n_B!)\\ &= -R\,x_A \ln x_A - R\,x_B \ln x_B\end{aligned}$$

A similar result is obtained from the thermodynamic considerations for an ideal solution, for which $\Delta H(\text{mixing}) = 0$:

$$\Delta S(\text{mixing}) = \frac{\Delta H(\text{mixing}) - \Delta G(\text{mixing})}{T}$$

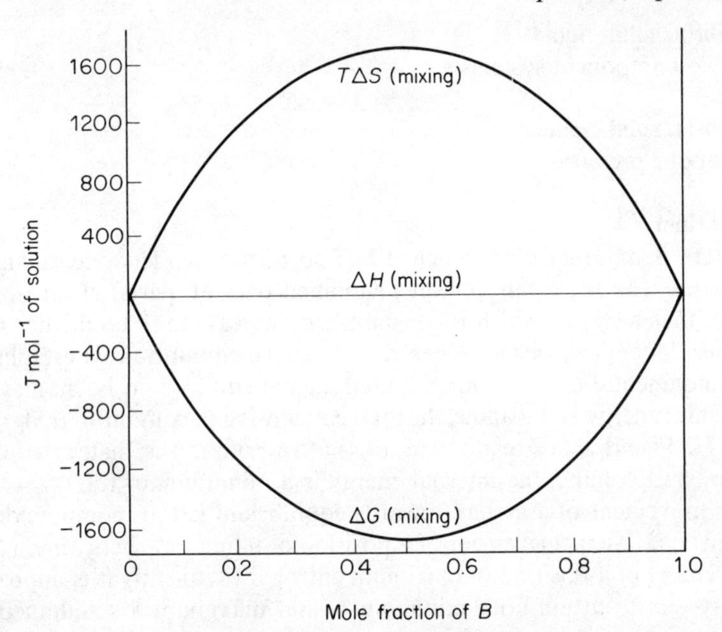

Figure E.8 The changes in the thermodynamic functions for the formation of 1 mole of an ideal solution at 298 K.

$$= -R\,x_A \ln x_A - R\,x_B \ln x_B$$

since the mole fractions x_A and $x_B < 1$, ΔS(mixing) > 0 and, for a given mole fraction, ΔS(mixing) is independent of the nature of the components and the temperature (figure E.8). ΔS(mixing) cannot be zero; thus, even at 0 K, the entropy of two pure substances separately is less than the entropy of any mixture of them.

Thus the driving force for the spontaneous mixing of two components is, in general, the entropy factor rather than an energy or an enthalpy factor.

See also Partition function; table A.I (p. 247); and Be, Da, Dic, E G H & R, I, K, L & R, Mo, P, Wy.

Equilibria, liquid–liquid

See Binary liquid mixture; Immiscible liquids; Partially miscible liquids.

Equilibria, liquid–vapour

See Clausius–Clapeyron equation; Vapour pressure.

Equilibria, solid–liquid
See Two-component systems.

Equilibria, solid–vapour
See Vapour pressure.

Equilibrium
A system is in equilibrium when it has no further tendency to change its properties. The total *entropy* (q.v.) (summed over all parts) of an isolated system, the energy of which is constant, will increase until no further spontaneous change can occur. When this happens, equilibrium is established. The fundamental criterion for thermodynamic equilibrium is: 'in a system of constant energy and volume, the total entropy is a maximum' or $(\partial S)_{U,V}=0$. Since U, V and S are related, an alternative criterion is that, at constant entropy and volume, the internal energy is a minimum, or $(\partial U)_{S,V}=0$.

The movement of a system towards equilibrium is thus compounded of two parts: (1) the achievement of a position of minimum energy and (2) the achievement of a position of maximum entropy. Frequently it is impossible for a system to attain both minimum U and maximum S simultaneously, and a compromise occurs. Thus a decrease in S may be allowable because the accompanying loss of energy outweighs the effect of the system becoming more ordered, and conversely a system may move to a position of higher energy provided this is offset by a sufficiently large increase in S.

More useful criteria for equilibrium under normal working conditions are: (1) at constant T and V, when only PV work is permitted, the *maximum work function*, A, is a minimum (see *Helmholtz free energy*), or $(\partial A)_{T,V}=0$ (A must be a minimum since U tends to decrease and S tends to increase in a system approaching equilibrium); and (2) at constant T and P, when only PV work is permitted, the *free energy* (q.v.), G, is a minimum, or $(\partial G)_{T,P}=0$.

The last criterion is probably the most useful under normal conditions, because the contributions from the *enthalpy* (q.v.) and free energy term can readily be seen from the equation

$$\Delta G=\Delta H-T\Delta S$$

Thus, if $(\partial G)_{T,P}<0$, the process (as written) is spontaneous; if $(\partial G)_{T,P}=0$, the process is in a state of equilibrium; and if $(\partial G)_{T,P}>0$, the process (as written) is not spontaneous.

At the macroscopic level, when equilibrium is attained, all activity apparently ceases, but at the microscopic level it is really the balance of opposing forces (e.g. in a chemical reaction the equality of the rates of the forward and back reactions).

The equilibrium of the component i between two phases α and β is established when the *chemical potential* (q.v.) of i is the same in both phases, since for equilibrium $dG = 0$ and, hence, $(\mu_i)_\alpha = (\mu_i)_\beta$.

See also Dnb.

Equilibrium constant

Many chemical reactions do not go to completion, but proceed to a position of *equilibrium* (q.v.) where reaction apparently ceases, often leaving considerable amounts of unchanged reactants. Under a standard set of conditions of T, P and concentration, the equilibrium position is always the same: i.e. there is a fixed relationship between the active masses of reactants and products. This is a dynamic state of equilibrium, in which the rate of formation of products from reactants is equal to the rate of formation of reactants from products. Thus, for the reaction

$$a\mathrm{A} + b\mathrm{B} \rightleftharpoons l\mathrm{L} + m\mathrm{M}$$

the *van't Hoff isotherm* (q.v.), relating the free energy change (see *free energy*) in the reaction to the active masses of reactants and products (designated [ ]) can be written:

$$\Delta G = \Delta G^\ominus + RT \ln \frac{[\mathrm{L}]^l [\mathrm{M}]^m}{[\mathrm{A}]^a [\mathrm{B}]^b}$$

But, for a system at equilibrium, $\Delta G = 0$; hence,

$$\Delta G^\ominus = -RT \ln \left\{\frac{[\mathrm{L}]^l [\mathrm{M}]^m}{[\mathrm{A}]^a [\mathrm{B}]^b}\right\}_e = -RT \ln K$$

where e indicates the active masses of the species at equilibrium.

Since G is a *state function* (q.v.), the ratio of the active masses of products and reactants is constant and independent of their actual individual values.

Hence, the equilibrium constant K is defined as

$$K = \left\{\frac{[\mathrm{L}]^l [\mathrm{M}]^m}{[\mathrm{A}]^a [\mathrm{B}]^b}\right\}_e = \prod_i [\mathrm{i}]^i = \exp(-\Delta G^\ominus/RT)$$

In actual practice the active mass is replaced by an *activity* (q.v.), concentration or pressure term. Table E.2 lists the recognised equilibrium constants.

For gas phase reactions, assuming ideal behaviour,

$$K_p = K_c(RT)^{\Delta\nu} = K_N(RT/N_\mathrm{A})^{\Delta\nu} = K_x\, P^{\Delta\nu}$$

Table E.2. List of equilibrium constants

Active mass	Equilibrium constant	SI unit
Partial pressure: p_i/N m^{-2}	$K_p = \prod_i p_i^{\,i}$	$(\mathrm{N\ m^{-2}})^{\Delta\nu}$
Relative partial pressure: $p_i/p^{\ominus}$	$K_{p/p^{\ominus}} = \prod_i (p_i/p^{\ominus})^i$	1
Concentration: c_i/mol m^{-3}	$K_c = \prod_i c_i^{\,i}$	$(\mathrm{mol\ m^{-3}})^{\Delta\nu}$
Molality: m_i/mol kg^{-1}	$K_m = \prod_i m_i^{\,i}$	$(\mathrm{mol\ kg^{-1}})^{\Delta\nu}$
Relative molality: $m/m^{\ominus}$	$K_{m/m^{\ominus}} = \prod_i (m_i/m^{\ominus})^i$	1
Mole fraction: x_i	$K_x = \prod_i x_i^{\,i}$	1
Molecules m^{-3}: N_i/m^{-3}	$K_N = \prod_i N_i^{\,i}$	$(\mathrm{molecules\ m^{-3}})^{\Delta\nu}$
Activity (for equilibrium in solution)	$K_{\mathrm{therm}} = \prod_i a_i^{\,i}$ $= \prod_i \gamma_i^{\,i}\, m_i^{\,i}$ $= K_{\mathrm{m}} \prod_i \gamma_i^{\,i}$	$(\mathrm{mol\ kg^{-1}})^{\Delta\nu}$

$\Delta\nu$ is the difference between the number of moles of products and reactants, and $\prod_i \gamma_i^i$ is the activity coefficient quotient to correct for non-ideal behaviour.

The thermodynamic equilibrium constant, K_{therm}, is a true constant for a reaction at a given temperature. On the other hand, the term $\prod_i \gamma_i^i$ is a quantity whose value for a gas reaction depends on the gases involved and the pressure, and, for reactions in solution, on the nature of the solvent and solute and the *ionic strength* (q.v.). Thus K_p, K_c and K_m are not true constants but vary with P or I, approaching the appropriate value of K_{therm} as $P \to 0$ or $I \to 0$.

Properties of equilibrium constants

(1) The equilibrium constant principle is only valid for systems at equilibrium, and unless true equilibrium concentrations or activities are used, the equations are not valid.

(2) The expression for K must always be written with the product-activities in the numerator; inversion of the ratio gives the constant for the reverse reaction, i.e. K(direct reaction) = $1/K$(reverse reaction).

(3) K for any reaction at fixed T is constant and independent of concentrations or pressures.

(4) K for a given reaction varies with T according to the *van't Hoff isochore* (q.v.).

(5) The magnitude of K determines the extent to which the reaction can proceed under given conditions. A large value for $K (\Delta G^\ominus \ll 0)$ indicates that the reaction favours the formation of products. A small value for $K (\Delta G^\ominus > 0)$ indicates that the reaction does not proceed to any appreciable extent under the given conditions (e.g. *dissociation constant*, q.v., of a weak acid).

As an example, consider the dissociation of water at 298 K and at a total pressure P; suppose that n molecules of $H_2O(g)$ are initially present and that α is the degree of dissociation:

$$H_2O(g) \rightleftharpoons H_2(g) + \tfrac{1}{2}O_2(g)$$

The number of molecules present:

	$H_2O(g)$	$H_2(g)$	$O_2(g)$	Total
Initially	n	0	0	
At equilib.	$n(1-\alpha)$	$n\alpha$	$\frac{1}{2}n\alpha$	$n(1+\frac{1}{2}\alpha)$

$$K_p = \frac{p(H_2)\,p^{1/2}(O_2)}{p(H_2O)} = \frac{\left(\dfrac{\alpha}{1+\alpha/2}\right)P\left(\dfrac{\alpha/2}{1+\alpha/2}\right)^{1/2} P^{1/2}}{\left(\dfrac{1-\alpha}{1+\alpha/2}\right)P}$$

$$= \frac{\alpha^{3/2}\,P^{1/2}}{\alpha^{1/2}(1-\alpha)(1+\alpha/2)^{1/2}}$$

Assuming that α is very small, this approximates to

$$K_p = \frac{\alpha^{3/2}\,P^{1/2}}{2^{1/2}}$$

$\Delta G_f^\ominus$ (H_2O, g. 298 K) $= -237.2$ kJ mol^{-1}; hence, $K_p = 2 \times 10^{-42}$ atm$^{1/2}$. P = saturation v.p. of water = 27 mmHg = 27/760 atm; thus $\alpha = 6.11 \times 10^{-28}$. The small extent of dissociation is due to the large free energy barrier.

Similarly, for the dissociation

$$N_2O_4(g) \rightleftharpoons 2NO_2(g)$$

$$K_p = \frac{p^2(NO_2)}{p(N_2O_4)} = \frac{[2\alpha P/(1+\alpha)]^2}{P(1-\alpha)/(1+\alpha)} = \frac{4\alpha^2 P}{1-\alpha^2}$$

whence,

$$\alpha = \left(\frac{K_p}{K_p + 4P}\right)^{1/2}$$

At a given T, although K_p remains constant, as the equilibrium pressure, P, becomes very high, α approaches 0, and as P approaches 0, so α approaches 1, as would be predicted by *Le Chatelier's principle* (q.v.).

(6) The qualitative effect of concentration of reactants and products and of adding inert gases on the position of a gaseous equilibrium can be predicted. For the general equilibrium,

$$K_p = \frac{x_L^l\, x_M^m}{x_A^a\, x_B^b} P^{\Delta\nu} = \frac{n_L^l\, n_M^m}{n_A^a\, n_B^b}\left(\frac{P}{\sum_i n_A}\right)^{\Delta\nu}$$

where $x_A = n_A/\sum_i n_A$.

On the addition of an inert gas at constant P ($\sum_i n_A$ is always increased) three conditions must be considered:

(a) $\Delta\nu = 0$, i.e. no change in the number of molecules on reaction, no change in the position of equilibrium.

(b) $\Delta\nu > 0$, $(P/\sum_i n_A)^{\Delta\nu}$ decreases; hence, for K_p to remain constant, the equilibrium shifts in the direction of the products (i.e. increase in n_L and n_M and decrease in n_A and n_B).

(c) $\Delta\nu < 0$, $(P/\sum_i n_A)^{\Delta\nu}$ increases and position of equilibrium shifts in the direction of the reactants.

The addition of an inert gas at constant volume causes no shift in the position of equilibrium, since P increases in proportion to $\sum_i n_A$.

The addition of reactants (products) at constant volume results in a shift in the position of equilibrium towards the products (reactants). At constant P the addition of reactants or products depends on the value of $\Delta\nu$, and each case must be considered individually.

For reactions in solution, the addition of reactants (products) displaces the equilibrium position towards products (reactants), so that K remains constant. The addition of a neutral electrolyte alters the ionic strength, I, and, hence, the activity coefficient correction term, and thus the position of equilibrium in an ionic reaction.

Determination of equilibrium constants

Chemical methods Chemical methods are not always possible, since any attempt to determine one species immediately disturbs the equilibrium. Chemical methods in which the equilibrium is frozen by suitable means are available:

(1) Direct chemical analysis for solubility product determination and for equilibria of the type

$$2Ag^+ + CaSO_4(s) \rightleftharpoons Ca^{2+} + Ag_2SO_4(s)$$

where various parts of the mixture can be separated before analysis.

(2) Distribution methods. The equilibrium

$$I_2 + I^- \rightleftharpoons I_3^-$$

can be studied by determining the distribution of iodine between water and an organic solvent (carbon tetrachloride) to give the distribution coefficient, $D = [I_2]_o/[I_2]_w$, where $[I_2]_o$ refers to the concentration in the organic phase and $[I_2]_w$ refers to the concentration in the water layer; and between aqueous KI and carbon tetrachloride (in which the ionic species are insoluble). Analysis of the aqueous phase gives $[\sum I_2]$ and the solvent phase gives $[I_2]_o$. Total $[I^-]$ is known; hence,

$$\begin{aligned}[\textstyle\sum I_2] &= [I_2]_w + [I_3^-] \\ &= [I_2]_o/D + [I_3^-] \\ [KI] &= [I^-] + [I_3^-]\end{aligned}$$

and all concentrations are available for the calculation of K. This method, in which concentrations are used in place of activities, is applicable to the dimerisation of weak acids in organic solvents and to the formation of complex ions.

(3) Freezing of the equilibrium mixture before analysis. In the gas phase equilibrium

$$H_2(g) + I_2(g) \rightleftharpoons 2HI(g)$$

at $T > 700$ K, equilibrium is rapidly attained and is not disturbed on rapid cooling to 300 K. At this temperature the concentrations of reactants and products can be determined without disturbing the equilibrium.

(4) Flow methods are available for homogeneous and heterogeneous gas equilibria, provided that the mixtures can be analysed after chilling.

Physical methods (1) For gaseous equilibria, for which $\Delta\nu \neq 0$, the measurement of the equilibrium pressure, P_e, enables the calculation of α and, hence, K_p. For the dissociation

$$A_2 \rightleftharpoons 2A$$

if m is the total mass of the mixture in a volume V, then the ideal pressure, assuming no dissociation, is

$$P_i = \frac{mRT}{MV}$$

where M is the molar mass of A_2. On dissociation, the measured equilibrium pressure is given by

$$P_e = (1+\alpha)P_i = (1+\alpha)\frac{mRT}{MV}$$

Thus, since m, R, T, V and M are known, α and, hence, K_p can be calculated. This method is suitable for studying the dissociation of N_2O_4, Cl_2, I_2, PCl_5.

(2) Conductometric methods for the determination of *dissociation constants* (q.v.) of acids and also hydrolysis constants.

(3) *Electrometric titrations*† for the determination of dissociation constants of acids and stability constants.

(4) *Galvanic cells*† for the determination of dissociation constants of acids, *solubility products* (q.v.) and the *ionic product of water* (q.v.).

(5) *Concentration cells*† of the type

⊖ Ag	$AgNO_3$ in presence of aq. ammonia $[a(Ag^+)]_1$	Saturated KNO_3	$AgNO_3$ $[a(Ag^+)]_2$	Ag ⊕

for which

$$E = \frac{RT}{F} \ln \frac{[a(Ag^+)]_2}{[a(Ag^+)]_1}$$

can be used to determine $[a(Ag^+)]_1$ and thus $[c(Ag^+)]_1$ and by difference $c(Ag(NH_3)_2^+)$ and $c(NH_3)$ and, hence, K_c for the equilibrium

$$Ag(NH_3)_2 \rightleftharpoons Ag^+ + 2NH_3$$

(6) Absorption spectrometry (in UV, visible and IR regions) provides a valuable method for the determination of the concentration of many molecules, ions and free radicals involved in dissociation equilibria in the gas phase (NO_2, I_2, etc.) and in the dissociation of acids in solution.

Calculation from other equilibrium constants (1) Using the *van't Hoff isochore* (q.v.), $K(T_2)$ for a reaction at T_2 can be calculated from a known value at another temperature, $K(T_1)$ and $\Delta H^\ominus$. Since $\Delta H^\ominus = f(T)$, the general form of the equation is

$$\ln K_p = -\frac{\Delta H_0^\ominus}{RT} + \frac{\Delta a}{R}\ln T + \frac{\Delta b}{2R}T + \frac{\Delta c}{6R}T^2 \ldots + J$$

where J, an integration constant, can be evaluated from a knowledge of K_p at one temperature.

(2) Use of simultaneous equilibria, e.g. in the water–gas equilibrium

$$H_2(g) + CO_2(g) \rightleftharpoons CO(g) + H_2O(g) \qquad K_1$$

the dissociation equilibria

$$H_2O(g) \rightleftharpoons H_2(g) + \tfrac{1}{2}O_2(g) \qquad K_2$$

and

$$CO_2(g) \rightleftharpoons CO(g) + \tfrac{1}{2}O_2(g) \qquad K_3$$

are also established, from which it can be shown that $K_1 = K_3/K_2$. Hence, a knowledge of K_2 and K_3 permits the calculation of K_1.

Calculation from thermodynamic data Any measurement of $\Delta G^\ominus$ must be very accurate, since the form of the relationship between K and $\Delta G^\ominus$ aggravates any error.

(1) $\Delta G^\ominus$ from tabulated entropy and enthalpy values.
(2) $\Delta G^\ominus$ for a given reaction can be obtained from tabulated values of standard free energy of formation or standard *electrode potentials*†.

Calculation from partition function For the general reaction

$$K = \frac{Q_L^l\, Q_M^m}{Q_A^a\, Q_B^b} \exp(-\Delta U_0/RT)$$

where ΔU_0 is the heat of reaction at 0 K, when both reactant and product molecules will be at their zero-point energy levels.

See also D & J, Dic, G & S, J & P, M & P.

Equipartition of energy
For an ideal monatomic gas, according to the kinetic theory of gases,

$$PV = \tfrac{1}{3}nm\bar{c}^2 = \tfrac{2}{3}\text{KE} = RT$$

where $\bar{c}^2$ is the mean square velocity of the molecules of mass m. Since the velocity and, hence, the kinetic energy of a molecule can be resolved into components along three axes mutually at right angles, it follows that each degree of freedom of translation (i.e. along each axis) contributes $\frac{1}{2}RT$ to the total energy, U. When the energy is fully exchanged among all the different degrees of freedom in a molecule (translational, rotational and vibrational), the same amount of energy, $\frac{1}{2}RT$, is associated with each degree of freedom—this is the principle of equipartition of energy. Since $C_V = (\partial U/\partial T)_V$, by using this principle it is possible to predict values of the *heat capacity* (q.v.) for various gases, assuming ideal behaviour.

A monatomic gas has only translational energy; hence, $C_V = \frac{3}{2}R = 24.94\ \mathrm{J\ K^{-1}\ mol^{-1}}$, a figure in agreement with the experimental values of C_V for He, Ne, Ar and Hg.

A molecule containing n atoms has $3n$ kinetic degrees of freedom, since each atom has freedom of movement in three dimensions; bonding forces, however, limit motion in certain directions to rotation or vibration instead of independent translation of the individual atoms. All molecules have three translational degrees of freedom. Non-linear molecules have three rotational degrees of freedom (since angular momentum has components along each of the three rotational axes). Linear molecules have two rotational degrees of freedom (all the atomic masses lie on one axis, and there is no moment of inertia about this axis). Each rotational degree of freedom contributes $\frac{1}{2}RT$ to the energy of the molecule. The remaining $(3n - 6)$ degrees of freedom for non-linear and $(3n - 5)$ for linear molecules are vibrational, one for each normal mode of deformation. Vibrational energy is partly kinetic and partly potential, so each vibration contributes $\frac{1}{2}RT$ for kinetic and $\frac{1}{2}RT$ for potential energy (i.e. total of RT per vibrational mode).

At low T, all molecules are in the vibrational ground state (collisions between molecules cannot provide sufficient energy for transition to a higher state). Vibration of molecules only occurs at high temperatures, since the quantum of vibration (*c.* 4000 J mol^{-1}) is large compared with the available energy $\frac{1}{2}RT$ (1200 J mol^{-1} at 300 K) per degree of freedom at lower temperatures. Most diatomic molecules have rotations fully in operation, sharing equally with translational motion at or above their b.p.s.

Thus for a diatomic (figure E.9) or linear polyatomic molecule at a temperature where no bending modes of vibration are active (i.e. rigid molecule),

$$U = U_0 + \tfrac{3}{2}RT + RT = U_0 + \tfrac{5}{2}RT$$

or

$$C_V = (\partial U/\partial T) = \tfrac{5}{2}R = 41.57\ \mathrm{J\ K^{-1}\ mol^{-1}}$$

Theoretical values of heat capacities for different types of gaseous molecules are given in table E.3.

For diatomic molecules, there is a steady increase in the experimental value of C_V with temperature, unlike the predicted curves. This is because

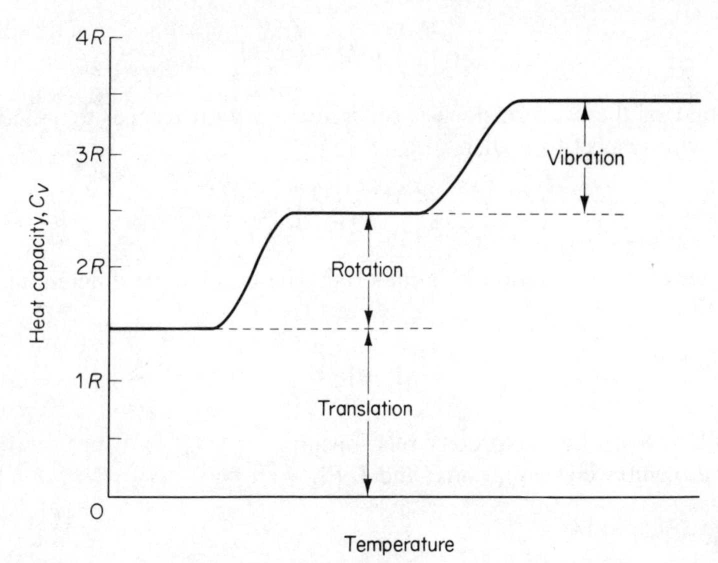

Figure E.9 Schematic diagram of the heat capacity of an ideal diatomic gas as a function of temperature.

Table E.3

Type of molecule	C_V	$\gamma = C_p/C_V$
Monatomic	$\frac{3}{2}R$	1.67
Rigid diatomic	$\frac{5}{2}R$	1.40
Rigid polyatomic, linear	$\frac{5}{2}R$	1.40
Rigid polyatomic, non-linear	$3R$	1.33
Vibrating diatomic	$\frac{7}{2}R$	1.29

all the molecules do not gain additional degrees of freedom at the same time; on average some possess more degrees of freedom than others.

Euler's reciprocity relationship

If X is an *exact differential* (q.v.) (e.g. *state function*, q.v.), i.e. $X = f(x, y)$, then

$$\mathrm{d}X = \left(\frac{\partial X}{\partial x}\right)_y \mathrm{d}x + \left(\frac{\partial X}{\partial y}\right)_x \mathrm{d}y = P\,\mathrm{d}x + Q\,\mathrm{d}y$$

Since dx and dy are independent, the coefficient of each must be the same in the two expressions, i.e.

Euler's reciprocity relationship

$$P = \left(\frac{\partial X}{\partial x}\right)_y; \quad Q = \left(\frac{\partial X}{\partial y}\right)_x$$

If the first of these expressions is differentiated with respect to y and the second with respect to x, then

$$\left(\frac{\partial P}{\partial y}\right)_x = \frac{\partial^2 X}{\partial x\, \partial y}; \left(\frac{\partial Q}{\partial x}\right)_y = \frac{\partial^2 X}{\partial x\, \partial y}$$

The order of differentiation is immaterial. The two partial differentials are equal; thus

$$\left(\frac{\partial P}{\partial y}\right)_x = \left(\frac{\partial Q}{\partial x}\right)_y$$

This is known as the reciprocity relationship.

See also Maxwell's equations; and I, P.

Eutectic arrest

Eutectic arrest is the duration of the constant temperature period at the eutectic temperature during the cooling of a melt of a *two-component con-*

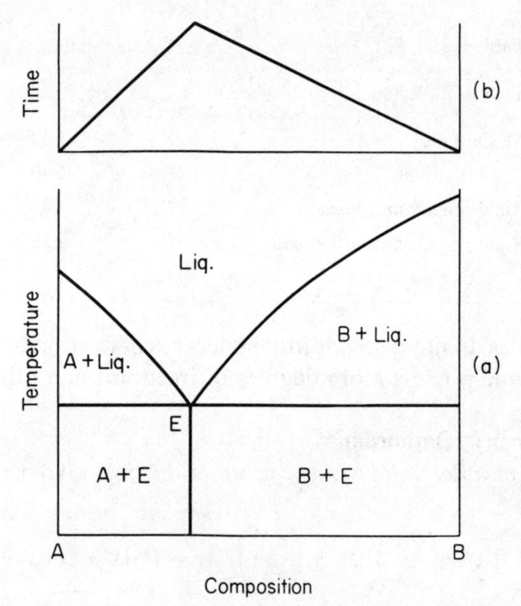

Figure E.10 (a) Phase diagram for two-component condensed system. (b) Duration of eutectic arrest.

densed system (q.v.). The arrest is a maximum for a melt having the eutectic composition (figure E.10). The composition of an unknown mixture can be determined non-destructively by plotting its cooling curve and comparing this with a calibration graph.

Eutectic mixture

In *two-component condensed systems* (q.v.) the solid mixture of constant proportion of components A and B which separates out from the liquid of the same composition at constant (eutectic) temperature is known as the eutectic mixture. Although each eutectic mixture has a fixed melting point, it is not a pure compound and cannot, in general, be represented by a simple empirical formula. Under a microscope two distinct crystalline forms are visible; these may be separated mechanically. It is often possible to selectively dissolve away one component.

The eutectic point is an invariant point, $f' = 0$ (see *invariant system*).

Exact differential

A *state function* (q.v.)X depends only on the state of the system and not on the way in which the system arrived at that state. The change in the value of the function in going from state 1 to state 2 depends only on the initial and final values of the function; i.e.

$$\Delta X = X_2 - X_1 = \int_{X_1}^{X_2} \mathrm{d}X$$

Integrands of this type are exact differentials. Integrals such as

$$\int_{x_1}^{x_2} x^2\, \mathrm{d}x = \tfrac{1}{3}x_2^3 - \tfrac{1}{3}x_1^3$$

are well defined and can easily be integrated, while others, such as

$$\int_{x_1, y_1}^{x_2, y_2} y\, \mathrm{d}x$$

cannot be evaluated without further information concerning the function $y(x)$. If it is known that y is a function of x, the graph of y against x can be drawn and the integral evaluated from the area under the curve; this is known as a line integral. A typical example in thermodynamics is the calculation of the work done during the expansion of a gas:

$$w = \int_1^2 đw = \int_{P_1 V_1}^{P_2, V_2} P\, \mathrm{d}V$$

which cannot be obtained, since the exact path must be known; the initial and final states are not sufficient, since P is a function not only of V but also of T, and T may change during the path of integration. The inexact differentials, $đw$ and $đq$, cannot be obtained by differentiation of a state function, and conversely $đw$ and $đq$ cannot be integrated to give w and q, respectively.

An important corollary of the fact that X is an exact differential is that

$$\mathrm{d}x = \left(\frac{\partial X}{\partial x}\right)_y \mathrm{d}x + \left(\frac{\partial X}{\partial y}\right)_x \mathrm{d}y$$

where x and y are the variables of state of the system (e.g. P, V and T) in terms of which X is expressed; e.g., since $U = f(V, T)$ it follows that

$$\mathrm{d}U = \left(\frac{\partial U}{\partial V}\right)_T \mathrm{d}V + \left(\frac{\partial U}{\partial T}\right)_V \mathrm{d}T$$

Exothermic process

An exothermic process, either chemical or physical, is one which is accompanied by the evolution of *heat* (q.v.), i.e. the heat content of the reactants is greater than that of the products. For example,

$$H_2(g) + \tfrac{1}{2}O_2(g) \longrightarrow H_2O(l) \qquad \Delta H(298\ \mathrm{K}) = -285.9\ \mathrm{kJ}$$

A reaction which is exothermic in one direction is endothermic in the reverse direction. The sign of ΔH is not a criterion of the spontaneity of the process.

See also Enthalpy.

Expansion

See Adiabatic process; Irreversible process; Isothermal process; Reversible process; Work.

Extensive property

See Property.

F

First law of thermodynamics

The first law of thermodynamics may be stated in various ways: (1) 'The energy of the Universe is a constant'. (2) 'The amount of energy of an isolated system remains constant although it may change from one form to another'.

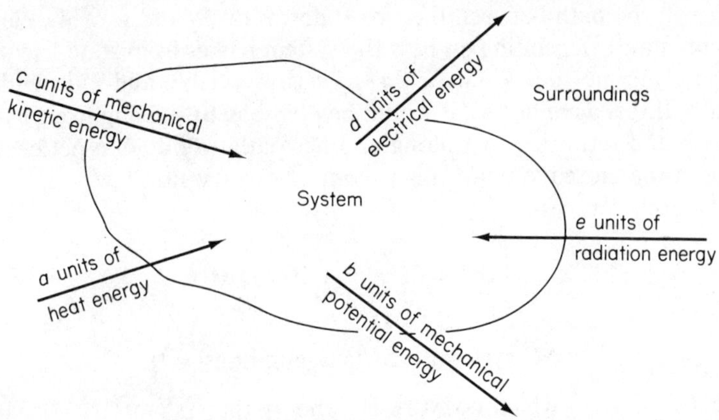

Figure F.1 Schematic illustration of different kinds of energy taken up or lost by a system in a given process.

(3) 'Energy can neither be created nor destroyed'. The interconvertibility of matter and energy, expressed by Einstein's equation ($\Delta U = mc^2$), shows that both should be combined in the more general statement that 'the total amount of energy, including that equivalent to mass, is constant in an isolated system'.

The first law is a law of nature and is consistent with the results of all known experiments; the large amount of observational material does not serve as proof but rather as support. The first law, unlike the *zeroth law of thermodynamics* (q.v.), is not limited in any way to states of equilibrium, and is regarded as having an unlimited range of application.

Schematically, the different kinds of energy that may be taken up or lost by a system when a process occurs are shown in figure F.1; thus, in passing from state 1 to state 2,

$$\Delta U = a - b + c - d + e$$

If the only interaction of the closed system with the surroundings is in the form of *heat* (q.v.) absorbed by the system, q, and the performance of *work* (q.v.), w, by the system, then the change in internal energy is

$$\Delta U = U_2 - U_1 = q - w$$

or, for an infinitesimal change,

$$\mathrm{d}U = đq - đw$$

The value of ΔU depends only on the final and initial values of U, and in no way on the path between the two states; q or $đq$ and w or $đw$ can have different values, depending on how the system passes from state 1 to state 2, but the algebraic sum $q - w$ or $đq - đw$ is invariable and independent of the path. If this were not so, it would be possible, by passing from 1 to 2 by one path and returning to 1 along another path, to obtain a net change of energy in the closed system; this is contrary to the first law.

For a cyclic process,

$$\oint \mathrm{d}U = \mathrm{d}U_{1\to 2} + \mathrm{d}U_{2\to 1} = 0$$

Also

$$\Delta U(\text{system}) + \Delta U(\text{surroundings}) = 0$$

For a process in which no work is done by the system, $\Delta U = q$; and for a process in which $q = 0$ (*adiabatic process*, q.v.), $\Delta U = -w$.

The first law is concerned only with changes in energy of material systems and not with the absolute energy of any system.

See also K, Ro.

Fractional distillation

See Distillation.

Free energy

Free energy or Gibbs energy, G (dimensions: m l^2 t^{-2}; units: J mol^{-1}), is a *state function* (q.v.) defined by the equation

$$G = H - TS = U - TS + PV \qquad \text{(F.1)}$$

For a pure substance, G depends on the chemical nature of the substance, the amount of substance in the sample, the state (solid, liquid or gas) in which it exists, the temperature and the applied pressure (for a solid or liquid remote from the critical point, the dependence on pressure can be neglected). The free energy of a mixture depends, in addition, on the composition of the mixture:

$$G = \mu_A n_A + \mu_B n_B \qquad \text{(F.2)}$$

where μ is the *chemical potential* (q.v.).

Equation (F.1) shows the compromise when both the *energy* (q.v.) and the *entropy* (q.v.) change, since in chemical systems of constant S, the equilibrium position is in the direction of lowest energy, while in systems of constant energy, the equilibrium position is in the direction of maximum

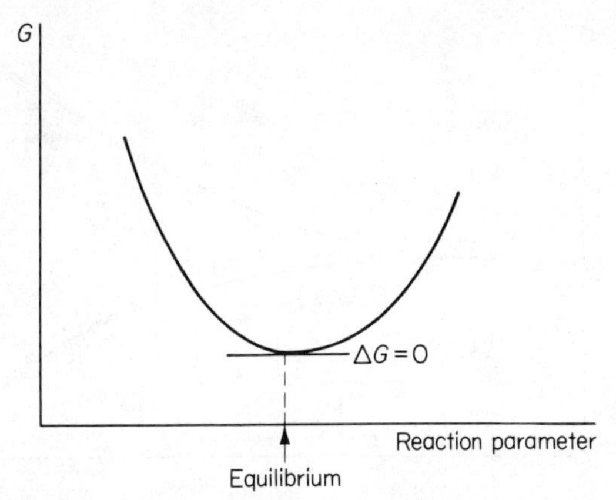

Figure F.2 Variation of free energy with the extent of a reaction.

entropy. At low temperatures, G depends mainly on the internal energy; at higher temperatures, the TS term becomes dominant.

The statement '$dG = 0$ for a reversible reaction at constant P and T at equilibrium' means that G is a maximum, a minimum or at a point of inflexion at equilibrium. Since $dG < 0$ for an irreversible process driving towards equilibrium, it follows that G is at a minimum value at equilibrium. Thus, under conditions of constant P and T, with no external work, G is the function to be minimised to find the conditions of equilibrium, i.e. $\Delta G = 0$ (figure F.2).

Although G is not directly measurable, free energy changes (see below) can be determined in a number of ways.

Differentiating equation (F.1) gives

$$dG = dU + PdV + VdP - TdS - SdT$$

which, if we introduce the equation for the *first law of thermodynamics* (q.v.) and the condition for a reversible change ($đq = TdS$), becomes

$$dG = VdP - SdT$$

whence,

$$\left(\frac{\partial G}{\partial T}\right)_P = -S \text{ and } \left(\frac{\partial G}{\partial P}\right)_T = V$$

The free energy of all substances decreases with increase in T; at 0 K, the free energy is equal to the total internal energy of a solid, which is equivalent to its *enthalpy* (q.v.), H.

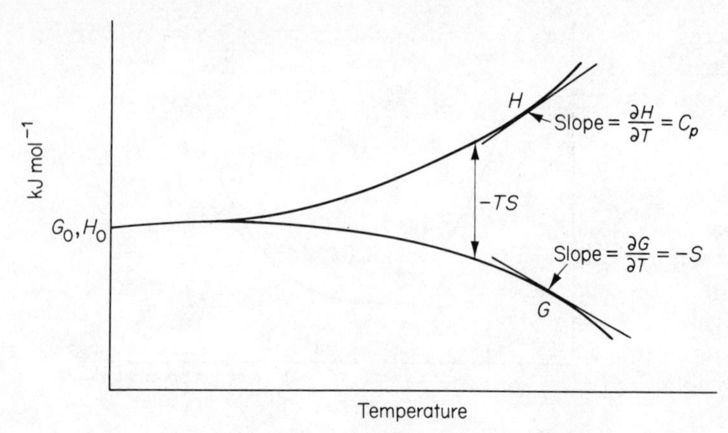

Figure F.3 Variation of free energy and enthalpy of a pure substance with temperature at constant pressure.

An increase in H accompanies an increase in temperature; the rate of increase is the *heat capacity* (q.v.) of the substance. At the same time G decreases with increase in temperature; the rate of decrease is the entropy of the substance (figure F.3).

An increase in pressure at constant temperature results in an increase in the free energy of a gas; for n mole of a perfect gas,

$$\left(\frac{\partial G}{\partial P}\right)_T = V = \frac{nRT}{P}$$

which on integration becomes

$$G = G^{\ominus} + nRT \ln (P/P^{\ominus})$$

where $G^{\ominus}$ is the standard free energy at reference pressure $P^{\ominus}$ and constant T; it is usual to choose $P^{\ominus}$ as unit pressure, when

$$G = G^{\ominus} + nRT \ln \frac{(P/\text{N m}^{-2})}{101\,325}$$

or, in general,

$$G = G^{\ominus} + RT \ln a$$

Thus $G^{\ominus}$, which is a function of temperature, is the free energy of the substance when present in its standard state (cf. *chemical potential*).

The most stable form of an element in the standard state is assigned a free energy of zero. It is thus possible to assign values to the standard free energies of formation (see below), $\Delta G_f^\ominus$, of compounds.

G is related to the *partition function* (q.v.):

$$G = -N_A kT \ln Q/N_A$$

Free energy change

The free energy change, ΔG (dimensions: m l^2 t^{-2}; units: J mol^{-1}), of a process or reaction is the maximum amount of useful *work* (q.v.) (e.g. electrical or mechanical work) which can be obtained by carrying out the process at constant temperature and pressure. Thus the free energy decrease, $-\Delta G$, represents the maximum net work, w', which the system can perform in a given change at constant temperature and pressure:

$$-\Delta G = -(\sum G(\text{final}) - \sum G(\text{initial})) = w_{\text{max}} - P\Delta V = w'$$

A process can only proceed spontaneously if the free energy is less at the end than at the start of the process; i.e. only if the system can accomplish work during the change. The free energy change of a process measures the over-all driving force, in terms of the compromise between the opposing effects of entropy and energy of the process:

$$\Delta G = \Delta H - T\Delta S \quad \text{(F.3)}$$

Thus, in general, for a constant pressure process or reaction, if $\Delta G < 0$, the process may proceed as written, i.e. it represents a spontaneous or natural process; if $\Delta G = 0$, the process will not proceed, i.e. it is at equilibrium; and if $\Delta G > 0$, the process will not proceed as written, but the reverse process may occur spontaneously, i.e. it represents a non-spontaneous or unnatural process. Note that in these statements the free energy change is that for the system; the role of the surroundings need not be included (cf. *entropy*).

The value of ΔG of a reaction can be considered as a measure of the available heat (i.e. available for performing useful work) as opposed to the total heat. The increase in order ($\Delta S < 0$) in a process requires work; this is manifested as non-functional $T\Delta S$ heat:

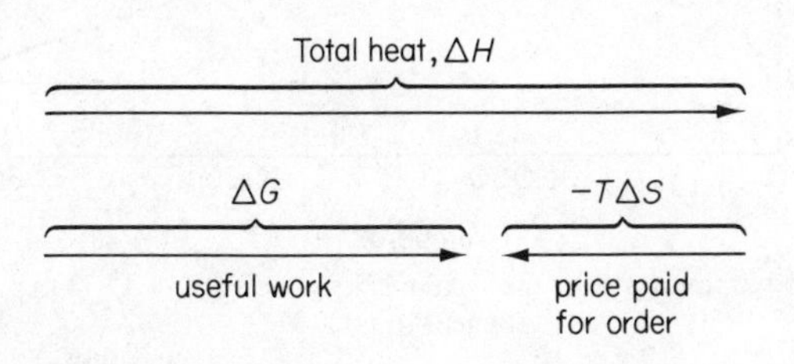

For such a process, the total heat evolved is still ΔH, but only the amount ΔG is available to do useful work. The greater the increase in ordering in such a process, the smaller the amount of heat available for useful work. In contrast, for a process in which the system becomes more disordered ($\Delta S > 0$), this disordering is available to do useful work:

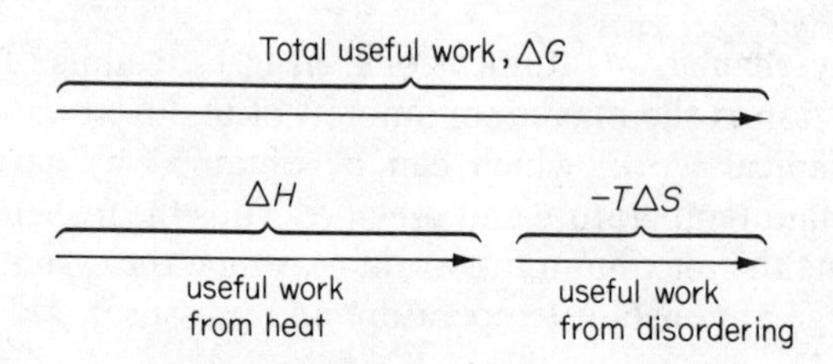

For isothermal changes in a system, the variation of ΔG with temperature is given by the *Gibbs–Helmholtz equation* (q.v.). Conversely, to evaluate ΔG as a function of temperature, the Gibbs–Helmholtz equation, in the form

$$\left(\frac{\partial(\Delta G/T)}{\partial T}\right)_P = -\frac{\Delta H}{T^2}$$

must be integrated. Since $H = f(T)$, *Kirchhoff's equation* (q.v.) must be introduced; and general integration gives

$$\Delta G = \Delta H_0 - \Delta a T \ln T - \tfrac{1}{2}\Delta b T^2 - \tfrac{1}{6}\Delta c T^3 \ldots + JT \tag{F.4}$$

where J is the integration constant, which can be evaluated if ΔH_0 and one value of ΔG are known. The empirical constants Δa, Δb, etc., are obtained

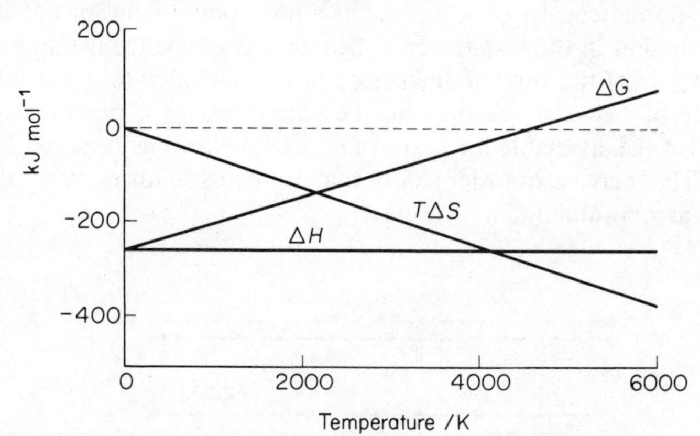

Figure F.4 Thermodynamic data for the system $H_2(g) + \frac{1}{2}O_2(g) \rightarrow H_2O(g)$ over the temperature range 0 to 6000 K.

from *heat capacity* (q.v.) data. Although tedious calculations cannot be avoided, these equations give near-linear ΔG–T plots, because the variations in the heat capacities of reactants and products are partly self-compensating in their effects on ΔH and ΔS.

Figure F.4 shows the variation of ΔG with T calculated from the appropriate values of $T\Delta S$ and ΔH for the reaction

$$H_2(g) + \tfrac{1}{2}O_2(g) \longrightarrow H_2O(g)$$

at 101 325 N m^{-2}. Over the entire temperature range, ΔH is almost constant (-242 kJ mol^{-1}), and apart from very low temperatures, ΔS is nearly constant (-44.3 J K^{-1} mol^{-1}), so that $T\Delta S$ becomes more negative with increasing temperature. As T tends to 0, $T\Delta S$ tends to 0 and ΔG is determined almost completely by ΔH, and the formation of water is highly feasible. At very high temperatures (> 4000 K), $T\Delta S$ has a larger negative value than ΔH, and so ΔG becomes positive and water vapour decomposes spontaneously into hydrogen and oxygen. This is an example of a reaction which, because of the negative value of ΔS, becomes less feasible with increase in T.

On the other hand, the industrial water gas reaction

$$C(s) + H_2O(g) \longrightarrow CO(g) + H_2(g)$$

is one which, because of a positive entropy change, becomes more feasible at higher temperatures (figure F.5).

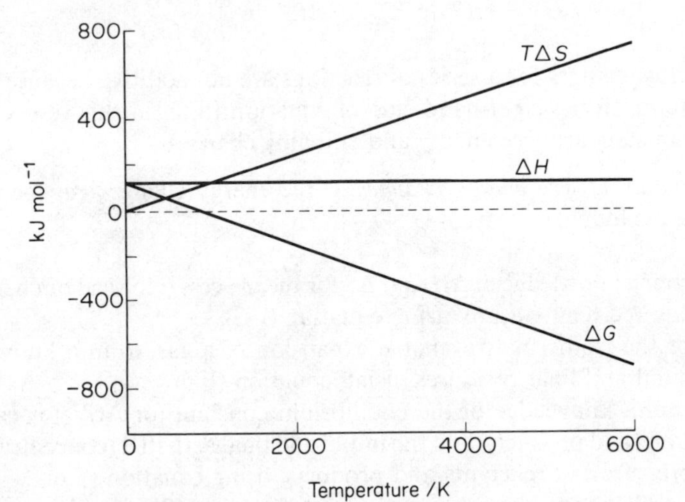

Figure F.5 Thermodynamic data for the system $C(s) + H_2O(g) \rightarrow CO(g) + H_2(g)$ over the temperature range 0 to 6000 K.

In both these examples ΔH and $T\Delta S$ have the same sign over the whole temperature range. In general, a low-energy state is highly ordered and, hence, a reaction resulting in greater order (disorder) for the system, with ΔS negative (positive), is almost always exothermic (endothermic), i.e. ΔH negative (positive). Ionic reactions in aqueous solution are an exception to this rule, owing to the organising of the solute ions on the water molecules.

The variation of ΔG with pressure at constant temperature is given by

$$\left(\frac{\partial(\Delta G)}{\partial P}\right)_T = V$$

Thus the isothermal free energy change, when n mole of an ideal gas is changed from pressure P_1 to P_2, is

$$\Delta G = G_2 - G_1 = nRT \ln P_2/P_1 \tag{F.5}$$

The free energy change in a chemical reaction is related to the *equilibrium constant* (q.v.) of the reaction and the initial and final activities (pressures or concentrations) of reactants and products, respectively, by the *van't Hoff isotherm* (q.v.), which, for the general reaction,

$$a\mathrm{A} + b\mathrm{B} \rightleftharpoons l\mathrm{L} + m\mathrm{M}$$

is

$$\Delta G = -RT \ln K_{\text{therm}} + RT \ln \frac{(a_\mathrm{L})^l\,(a_\mathrm{M})^m}{(a_\mathrm{A})^a\,(a_\mathrm{B})^b} \tag{F.6}$$

Free energy changes for a series of reactions are not additive because of the second term on the right-hand side of equation (F.6); in this respect they differ from standard free energy and enthalpy changes.

Determination of free energy changes Free energy changes can be determined in the following ways:

(1) From a knowledge of ΔH and ΔS for the process (physical or chemical) at the required temperature, using equation (F.3).

(2) For the isothermal reversible expansion of a gas, from a knowledge of the initial and final pressures, using equation (F.5).

(3) From a knowledge of the equilibrium constant (or $\Delta G_\mathrm{f}^\ominus$ for each of the reactants and products) and the initial and final activities (concentrations, partial pressures) of reactants and products, using equation (F.6).

(4) For reactions occurring in a reversible *galvanic cell*†, from a knowledge of the e.m.f., E:

$$\Delta G = -nFE$$

(5) The free energy change at different temperatures may be evaluated from equation (F.4).

Standard free energy change

The standard free energy change, $\Delta G^\ominus$ (units: J mol^{-1}), is the change in free energy accompanying the conversion of reactants in their standard states to products in their standard states. Standard free energies are additive:

$$\Delta G^\ominus = \sum G_f^\ominus(\text{products}) - \sum G_f^\ominus(\text{reactants})$$

Thus, consider the calculation of the standard free energy of combustion of ethanol at 298 K:

$$C_2H_5OH(l) + 2\tfrac{1}{2}O_2(g) \longrightarrow 2CO_2(g) + 3H_2O(l)$$
$$\Delta G^\ominus = 2(-394.383) + 3(-237.192) - (-176.766) = -1332.596 \text{ kJ}$$

i.e. this is a thermodynamically feasible process.

Standard free energy changes for reactions calculated by this method are of great use in determining the thermodynamic feasibility of reactions. For slow reactions, a catalyst should be sought only if the equilibrium is favourable. Thus, in the formation of CS_2 from methane and sulphur,

$$CH_4(g) + 4S(s) \longrightarrow CS_2(l) + 2H_2S(g)$$
$$\Delta G^\ominus(298 \text{ K}) = 63.60 + 2(-33.02) - (-50.71) = 48.27 \text{ kJ}$$

The large increase in G accompanying this process indicates that the amount of liquid CS_2 at equilibrium would be small, no matter what the catalyst. In contrast, a suitable catalyst at 298 K would permit the reaction

$$2CO(g) \longrightarrow C(s) + CO_2(g)$$
$$\Delta G^\ominus(298 \text{ K}) = -394.38 - 2(-137.26) = -119.86 \text{ kJ}$$

For this reaction, a similar calculation shows that $\Delta S^\ominus(298 \text{ K}) = -176.47$ J K^{-1}. Thus $\Delta G^\ominus$ will become less negative as the temperature rises; the temperature at which $\Delta G^\ominus = 0$ (i.e. the temperature above which the reverse reaction becomes favoured) can be approximately calculated; if $\Delta S^\ominus$ and $\Delta H^\ominus$ are assumed to be independent of temperature, then, according to the equation

$$\Delta G^\ominus(T) = \Delta H^\ominus(298 \text{ K}) + \Delta S^\ominus(298 \text{ K})[298 - T]$$
$$= -119\,860 - 176.47[298 - T]$$

$\Delta G^\ominus(T) = 0$ when $T = 977$ K. A more accurate value can be obtained by using heat capacity data to determine $\Delta H^\ominus$ as a function of temperature.

Free energy

For a system in equilibrium at constant temperature and pressure $\Delta G = 0$; applying this restriction to the van't Hoff isotherm (equation F.6), it follows that

$$\Delta G^{\ominus} = -RT \ln \left[\frac{(a_L)^l (a_M)^m}{(a_A)^a (a_B)^b} \right]_e = -RT \ln K_{therm}$$

Evidently K_{therm} $[= \exp(-\Delta G^{\ominus}/RT)]$ must be a constant, since $\Delta G^{\ominus}$ has a fixed value at a given temperature. Hence, K_{therm} can be calculated from a knowledge of $\Delta G^{\ominus}$ or vice versa. For an isothermal, isobaric equilibrium between ideal gases, this equation becomes

$$\Delta G^{\ominus} = -RT \ln \left[\frac{(p_L)^l (p_M)^m}{(p_A)^a (p_B)^b} \right]_e = -RT \ln K_p$$

Thus, from a knowledge of the equilibrium constant or $\Delta G^{\ominus}$ at a given temperature, the over-all free energy change, ΔG, for a non-equilibrium reaction can be evaluated for specified initial and final activities (pressures).

$\Delta G^{\ominus}$ varies with temperature according to the equation

$$\Delta G^{\ominus} = \Delta H_0^{\ominus} - \Delta a T \ln T - \tfrac{1}{2}\Delta b T^2 - \tfrac{1}{6}\Delta c T^3 \ldots + JT$$

where the heat capacity values and the enthalpy change are those for gases, extrapolated to zero pressure.

With the increasing use of spectroscopic data, an alternative method of presentation of free energy data has come into use. This consists of the tabulation of the *free energy function* (q.v.) $(G_T^{\ominus} - H_0^{\ominus})/T$; this method avoids the use of empirical equations. From these tables, standard free energies of formation, $\Delta G_f^{\ominus}$, can be evaluated for each of the reactants and products at the required temperature and, hence, $\Delta G^{\ominus}$ for the reaction (see below).

Standard free energy of formation

The standard free energy change of formation, $\Delta G_f^{\ominus}$, is the free energy change accompanying the formation of a substance, in its standard state, from its elements in their standard states, at the specified temperature. Thus the standard free energy of formation of liquid water at 298 K refers to the reaction

$$H_2(\text{g, 1 atm}) + \tfrac{1}{2}O_2(\text{g, 1 atm}) \longrightarrow H_2O(\text{l, 1 atm})$$

for which

$$\Delta G = \Delta G_f^{\ominus}(H_2O, \text{l}, 298\text{ K}) = -237.192\text{ kJ mol}^{-1}$$

Typical values for $\Delta G_f^{\ominus}$ are listed in table A.I (p. 247).

Compounds which have a large positive value for $\Delta G_f^\ominus$ are thermodynamically unstable with respect to their elements. Synthesis of such compounds may be very difficult, so suitable reactants must be found to give an over-all free energy decrease. Thus the formation of benzene by the polymerisation of acetylene (also unstable with respect to its elements) is highly favoured:

$$3C_2H_2(g) \longrightarrow C_6H_6(l)$$
$$\Delta G^\ominus = 124.5 - 3 \times 209.2 = -503.1 \text{ kJ}$$

Extrapolation of free energy equations shows that benzene becomes thermodynamically unstable with respect to acetylene above 1800 K, while acetylene becomes stable with respect to its elements above 4000 K.

It is possible to calculate $\Delta G_f^\ominus$ from tabulated values of the *free energy function* (q.v.); such a method avoids the use of empirical equations to calculate $\Delta G_f^\ominus$ at temperatures other than that specified. Thus, for the formation of a compound (from its elements),

$$\frac{\Delta G_f^\ominus}{T}(T,\text{K}) = \left\{\frac{\Delta H_{f0}^\ominus}{T} + \frac{(G_T^\ominus - H_0^\ominus)}{T}\right\}_{\text{Compound}} - \sum\left(\frac{G_T^\ominus - H_0^\ominus}{T}\right)_{\text{Element}}$$

Thus, from a knowledge of the free energy of formation of each substance in a reaction, the over-all standard free energy change of the reaction can be evaluated.

Standard free energy of an ion

The standard free energy of an ion is the partial molar free energy of the ion in solution, relative to the conventional reference standard free energy of the hydrogen ion equal to zero, $\bar{G}^\ominus[H_3O^+, a(H_3O^+) = 1] = 0$. Consider the reaction

$$Cd(s) + 2H^+(aq) \longrightarrow Cd^{2+}(aq) + H_2(g)$$

which occurs in the cell

$$\text{Cd} \,\big|\, Cd^{2+}(aq) \,\vdots\, H^+(aq) \,\big|\, H_2(g), \text{Pt}$$

If all the reactants are in their standard states

$$-nFE^\ominus = \Delta G^\ominus = \bar{G}^\ominus(Cd^{2+}, aq) + \bar{G}^\ominus(H_2, g) - \bar{G}^\ominus(Cd, s) - 2\bar{G}^\ominus(H_3O^+, aq)$$

$E^\ominus = 0.402$ V and $\bar{G}^\ominus(Cd, s) = \bar{G}^\ominus(H_2, g) = \bar{G}^\ominus(H_3O^+, aq) = 0$ and since the elements are in their standard states, it follows that

$$\bar{G}^\ominus(Cd^{2+}, aq) = -2 \times 96\,477 \times 0.402 = -77.568 \text{ kJ mol}^{-1}$$

From such tabulated data (table A.I, p. 247), it is possible to calculate standard free energy changes in ionic reactions.

Free energy of mixing
The free energy of mixing, for a mixture of two ideal components, is

$$\Delta G(\text{mixing}) = x_A RT \ln x_A + x_B RT \ln x_B$$

ΔG(mixing) is always less than zero, and, for a given mole fraction, is independent of the nature of the components. For *real solutions* (q.v.) ΔG(mixing) is greater or less than that for an *ideal solution* (q.v.), depending on the deviation from *Raoult's law* (q.v.).

See also Bar, Da, Dic, G & S, I, K, L & R, Mo, Wal, War, Was.

Free energy function
The free energy function, $(G_T^\ominus - H_0^\ominus)/T$ (units: J K^{-1} mol^{-1}), for ideal gases is obtained from the equation

$$(G_T^\ominus - H_0^\ominus)/T = 30.472 - 19.144\{\tfrac{3}{2}\log M + \tfrac{5}{2}\log T\} - 19.147 \log Q$$

Tabulated values (table A.VI, p. 254), which have been obtained from spectroscopic measurements, are of great use in the calculation of $\Delta G_f^\ominus$ values and, hence, $\Delta G^\ominus$ for reactions at temperatures other than that specified. The advantage of the method is that it avoids the use of empirical *heat capacity* (q.v.) equations.

See also Free energy; Partition function; and K, L & R, Nat. Bur. Standards, Circular 500 (1952).

Freezing point
The normal f.p. is the temperature at which solid and liquid forms of a substance have the same v.p. The direct consequence of the lowering of the v.p. of a volatile solvent by a non-volatile solute is that the f.p. of the solution is lower than that of the pure solvent (figure F.6). The solid solvent (whose v.p. is unaffected by the presence of the solute) is thus in equilibrium with the liquid solution at a lower temperature than that of its equilibrium with pure liquid solvent. The theory is only applicable to the case when solvent crystals separate out from the solution. The depression of f.p., ΔT_c, of the solvent A is related to the lowering of the v.p., and, hence, by *Raoult's law* (q.v.) to the concentration of solute B in solution:

$$\Delta T_c = \frac{RT_0^2}{L_f} \ln p^\ominus/p = \frac{RT_0^2}{L_f} x_B$$

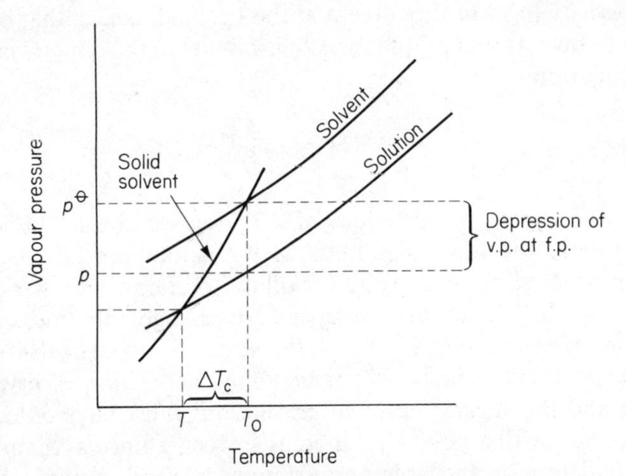

Figure F.6 The vapour pressure–temperature curves for solvent and solution near the freezing point of the solvent.

It is customary to use the molality of B, represented by m_B, rather than x_B; if n_A is the number of moles of solvent in 1 kg solvent, then

$$\Delta T_c = \left(\frac{RT_0^2}{n_A L_f}\right) m_B = k_c m_B$$

where the expression in parentheses is the molal cryoscopic constant, k_c, for the given solvent.

The freezing point depression is clearly a function of the properties of the solvent, and is independent of any feature of the solute except its true concentration. Depression of f.p. is thus a *colligative property* (q.v.).

Table F.1. Molal cryoscopic constants for various solvents

Solvent	f.p./ °C	k_c/K kg mol^{-1}
Acetic acid	16.6	3.9
Benzene	5.5	5.11
Bromoform	7.8	14.3
Camphor*	173	40
Cyclohexane	6.5	20.2
Nitrobenzene	5.3	6.9
Water	0	1.86

* Since commercial camphor is not usually pure, k_c should be determined for each sample by using a solute of known molecular mass.

The *activity* (q.v.) of the solvent at the f.p., and, hence, that of the solute, can be obtained from f.p. studies. ΔT_c is related to the *osmotic pressure* (q.v.) by the equation

$$\Pi V = L_f \frac{\Delta T_c}{T_0}$$

From the experimental values of ΔT_c for solutions of known weight concentration, the molecular mass of the solute can be calculated. The equations quoted are only valid for dilute solutions, and, hence, values of ΔT_c are small. These small temperature changes are measured with a *Beckmann thermometer* (q.v.) or with *thermocouples* (q.v.) or *thermistors* (q.v.). In the experimental methods, supercooling, deposition of crystals on the detector and the separation of solid solutions must be avoided.

If camphor, with a large k_c, is used as solvent, a normal thermometer can be used to determine the large depression of f.p.; the accuracy of this method, due to Rast, is low, but is adequate for some applications in organic chemistry.

The experimental methods are limited to compounds of low relative molecular mass; otherwise ΔT_c is too small to be measured accurately. Anomalous values of $M_r(B)$ are obtained for solutes which undergo association or dissociation in the solvent. For example, benzoic acid in acetone gives a value of ΔT_c corresponding to a molecular mass of 122, while in benzene it corresponds to a molecular mass of 244; this is because benzoic acid dimerises in benzene. Dissociation of the solute into ions gives values of $M_r(B)$ which are a fraction of the expected value, depending on the number of ions formed; each ion acts to inhibit the escape of solvent molecules, and, hence, gives rise to an independent lowering of v.p. and f.p.

See also Boiling point; and Bar, G & S, J & P.

Freezing point depression
See Freezing point.

Fugacity
Fugacity, P^* or p^* (units: $N\,m^{-2}$; atm), is a function introduced by G.N. Lewis to faciliate the application of thermodynamics to real systems. Thus, when fugacities are substituted for partial pressures in the equilibrium constant expression, which applies strictly only to ideal systems, a true equilibrium constant results for real systems as well.

For real gases,

$$\bar{V} \neq RT/P$$

and, hence, the *chemical potential* (q.v.)

$$\mu \neq \mu^{\ominus} + RT \ln P$$

This difficulty may be overcome either by substituting for P a more complicated expression derived from an empirical equation of state (e.g. *van der Waals' equation*, q.v.) or by changing the variable P so that real gas behaviour is predicted. The new variable, 'fugacity' (P^* for the single imperfect gas, p_i^* for the partial fugacity of a component of a mixture of imperfect gases), which has the same dimensions and general properties as pressure, makes the ideal gas equation valid for real gases. Since all gases tend to ideal behaviour as P approaches zero, fugacity is defined by

$$\mu = \mu^{\ominus} + RT \ln P^* \qquad \text{(F.7)}$$

and

$$\lim_{P \to 0} P^*/P = 1$$

Thus the free energy change in an isothermal process, $RT \ln (P_2^*/P_1^*)$, approaches that for an ideal gas, $RT \ln (P_2/P_1)$, as P approaches zero. This definition of fugacity applies for the solid and liquid states as well as for the gaseous state.

The fugacity, which cannot be measured directly, is most readily calculated from P, V, T data. If α is the difference between the molar volume of an ideal gas (RT/P) and the true measured molar volume, $\bar{V}$, under the same conditions (figure F.7),

$$\bar{V} = RT/P - \alpha \qquad \text{(F.8)}$$

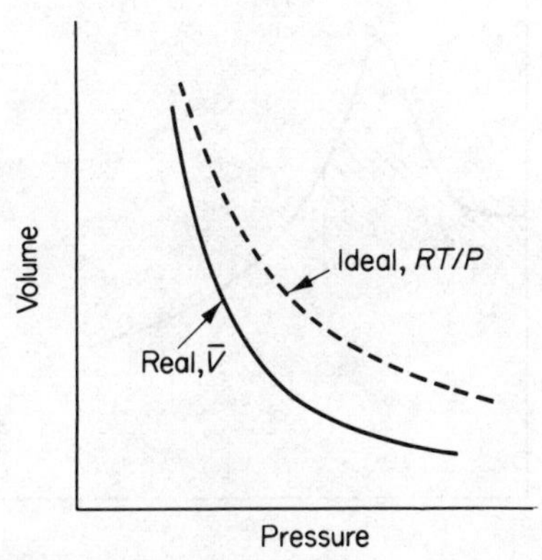

Figure F.7 P/V isothermals for a real and an ideal gas.

Hence,

$$d\mu = \bar{V}dP = RT \, d \ln P^* \tag{F.9}$$

Combining equations (F.8) and (F.9) and rearranging gives

$$\int_0^{P^*} d \ln P^* = \int_0^P d \ln P - \frac{1}{RT}\int_0^P \alpha \, dP$$

at $P = 0$, $P^* = P$; hence,

$$\ln P^* = \ln P - \frac{1}{RT}\int_0^P \alpha \, dP \tag{F.10}$$

The integral of $\alpha \, dP$ is evaluated graphically from a determination of α over a wide range of pressures and extrapolation to zero pressure (figure F.8).

The behaviour of most pure gases can be represented by a single graph of the compressibility factor $Z (= PV/RT)$. If Z is plotted against the reduced pressure $P_r (= P/P_c)$ at a given reduced temperature $T_r (= T/T_c)$, where P_c and T_c are the critical pressure and temperature, respectively, then all gases fit a simple curve. Z is related to α:

$$\alpha = \frac{RT}{P}(1 - Z)$$

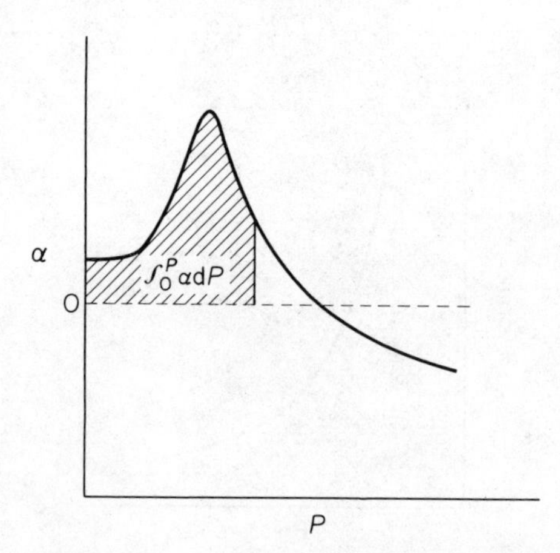

Figure F.8 Graph for the integration of αdP.

Hence,

$$\ln \frac{P^*}{P} = -\int_0^P (1 - Z) \frac{dP}{P} \tag{F.11}$$

This integration can be carried out to give graphs of the fugacity coefficient or activity coefficient, $f = P^*/P$, as a function of P_r and T_r. A single set of graphs (figure F.9) is applicable to all gases within the precision to which the compressibility factor chart is valid.

f varies with P. At low values of P, $f < 1$, because the molecules are so far apart that bulk is not a major factor determining their motion. At high P, the bulk becomes important; this is related to the b term of the van der Waals' equation and at very high $P, f > 1$.

Fugacity is a measure of the escaping tendency of real gases from the condensed phase. Since at *equilibrium* (q.v.) the chemical potential of a component in the liquid must be equal to that of the component in the gas phase, the fugacities of one component in two phases must be equal. In a multi-component system, the escaping tendency and, hence, the fugacity,

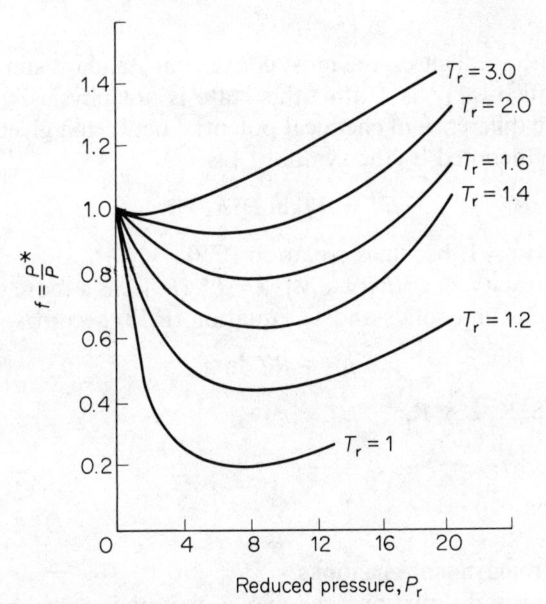

Figure F.9 Variation of activity coefficient of a gas with reduced pressure at various reduced temperatures.

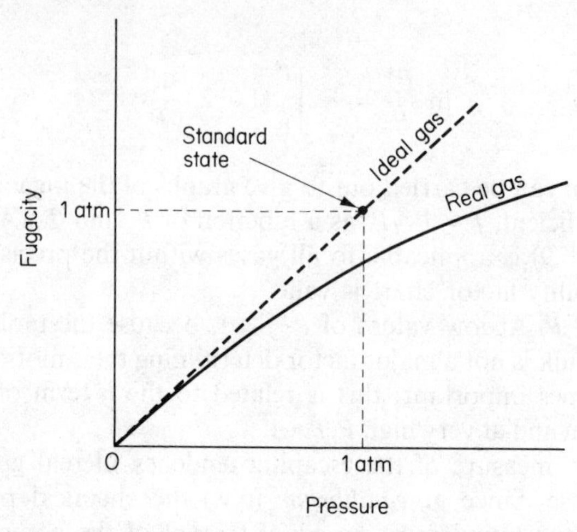

Figure F.10 Definition of standard state of unit fugacity.

p_i^*, is less than if it were present in the pure form, $(p_i^{*\cdot})$:

$$p_i^* = x_i p_i^{*\cdot} \tag{F.12}$$

By analogy with ideal gases, the most convenient standard state for a real gas is one in which $(p_i^*)^\ominus$ is 1 atm; this state is not physically realisable (figure F.10). The difference in chemical potential between a given state and a reference state (denoted by the symbol $^\ominus$) is

$$\mu - \mu^\ominus = RT \ln P^*/(P^*)^\ominus \tag{F.13}$$

which, when $(P^*)^\ominus = 1$, becomes equation (F.7).

The relative fugacity, or *activity* (q.v.), $a = P^*/(P^*)^\ominus$, is a more convenient function for liquids and solids and so equation (F.13) becomes

$$\mu = \mu^\ominus + RT \ln a \tag{F.14}$$

See also G & S, K, L & R.

Function
See State function.

Fundamental thermodynamic equations
The four fundamental equations relating a thermodynamic or *chemical potential* (q.v.) to its appropriate independent variables are:

$$\begin{aligned}
\mathrm{d}U &= \quad T\mathrm{d}S - P\mathrm{d}V + \sum_i \mu_i \, \mathrm{d}n_i \\
\mathrm{d}A &= -S\mathrm{d}T - P\mathrm{d}V + \sum_i \mu_i \, \mathrm{d}n_i \\
\mathrm{d}H &= \quad T\mathrm{d}S + V\mathrm{d}P + \sum_i \mu_i \, \mathrm{d}n_i \\
\mathrm{d}G &= -S\mathrm{d}T + V\mathrm{d}P + \sum_i \mu_i \, \mathrm{d}n_i
\end{aligned}$$

By means of these equations, all the thermodynamic functions U, H, A and G can be expressed in terms of a chosen chemical potential and its derivatives with respect to the corresponding independent variables S, T, V and P. For example, defining

$$G = f(T, P, n_i)$$

then

$$\begin{aligned}
S &= -(\partial G/\partial T)_{P,n_i} \\
H &= G - T(\partial G/\partial T)_{P,n_i} \\
V &= (\partial G/\partial P)_{T,n_i} \\
A &= G - P(\partial G/\partial P)_{T,n_i} \\
U &= G - T(\partial G/\partial T)_{P,n_i} - P(\partial G/\partial P)_{T,n_i} \\
\mu_i &= (\partial G/\partial n_i)_{T,P}
\end{aligned}$$

Similarly, all the functions can be expressed in terms of U and its differential coefficients with respect to S, V and n_i.

G

Gas constant

The gas constant, R (dimensions: m l^2 t^{-2} deg^{-1}; units: J K^{-1} mol^{-1}), is related to the pressure, volume and temperature of an ideal gas by

$$R = PV/nT$$

At 273.15 K and 101 325 N m^{-2} pressure, 1 mole of an ideal gas occupies a volume of 22.414×10^{-3} m^3; hence

$$\begin{aligned}
R &= \frac{101\,325 \times 22.414 \times 10^{-3}}{273.15} \text{ N m K}^{-1}\text{ mol}^{-1} \\
&= 8.314 \text{ J K}^{-1}\text{ mol}^{-1}
\end{aligned}$$

Gibbs adsorption isotherm

The possible effects of a dissolved substance on the *surface tension* (q.v.), γ, of the solvent are illustrated in figure G.1. In solutions of type 1, addition

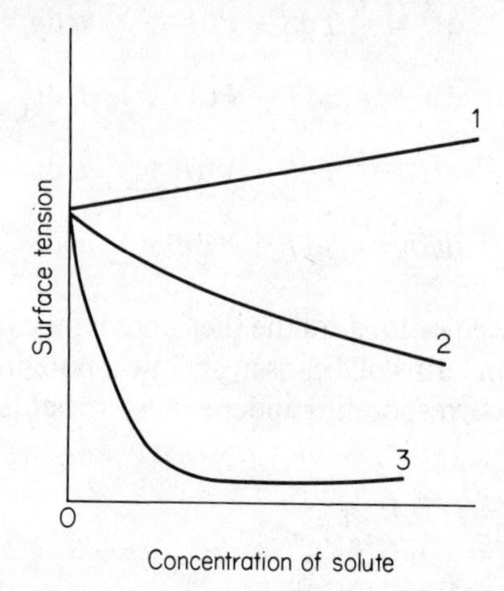

Figure G.1 Variation of the surface tension of a solution with concentration.

of the solute leads to a small increase in γ; such solutes, referred to as 'surface-inactive agents', include inorganic electrolytes and sucrose in water, and aniline in cyclohexane. In solutions of types 2 and 3, addition of the solute leads to a gradual decrease in γ; such solutes, referred to as 'surface-active agents', are generally long-chain compounds of the type $C_nH_{2n+1}X$ (where X is carboxyl, sulphonate, alcohol, amide, amine, etc.) in water. For a given homologous series, the depression of γ at a given concentration increases with the value of n (i.e. type 2 refers to short-chain, and type 3 to longer-chain compounds, up to $n = 10$–12).

This surface activity is due to the unequal distribution of solute between the surface phase and the bulk solution. From a consideration of the free energy (at constant P and T) of the surface and of the bulk phases in terms of the variables of concentration and surface area, Gibbs proved that, at equilibrium for solute B,

$$\Gamma_B = -\frac{1}{RT}\left(\frac{\partial \gamma}{\partial \ln a_B}\right)_{T,P}$$

$$\approx -\frac{c_B}{RT}\left(\frac{\partial \gamma}{\partial c_B}\right)_{T,P}$$

where c_B is the bulk concentration of B and Γ_B the surface excess concentration of B per unit area of surface. Γ_B is really the concentration of B adsorbed at the interface.

This equation, the Gibbs adsorption isotherm, shows that a substance

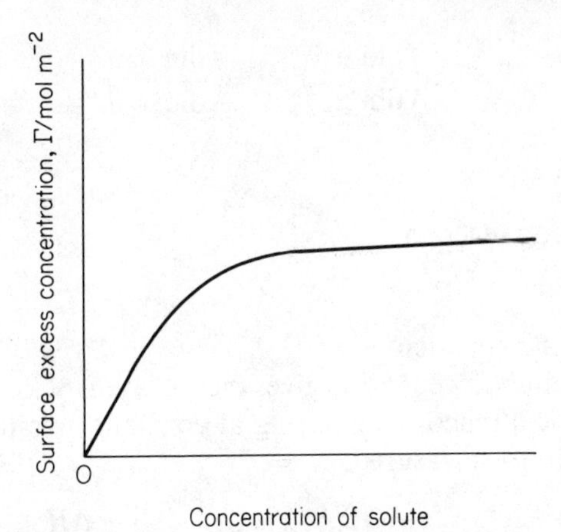

Figure G.2 Adsorption of a surface active solute at an air–water interface.

which lowers γ accumulates in the interface and that Γ_B is positive. A substance which produces an increase in γ is less concentrated in the surface than in the bulk, and Γ_B is negative. The isotherm, which has been verified experimentally, is of use in calculating the area per molecule of solute at an interface from measurements of γ at different concentrations (figure G.2).

Gibbs–Duhem equation

The Gibbs–Duhem equation relates the activities of the components of a liquid pair and is useful in calculating the *activity* (q.v.) of one component (solvent) from that of the other (solute). For a binary liquid mixture,

$$n_A d\bar{J}_A + n_B d\bar{J}_B = 0$$

where $\bar{J}_A$ and $\bar{J}_B$ are the *partial molar quantities* of A and B. Since the activity is related to the partial molar *free energy* (q.v.), $\bar{G}$, or the *chemical potential* (q.v.), μ, the equation may be written (now in terms of mole fractions)

$$x_A d\bar{G}_A + x_B d\bar{G}_B = 0 \text{ or } x_A d\mu_A + x_B d\mu_B = 0$$

Since

$$\mu_A = \mu_A^{\ominus} + RT \ln a_A$$

it follows that

$$x_A d\ln a_A + x_B d\ln a_B = 0$$

Since $x_A = 1 - x_B$, therefore $dx_A = -dx_B$ and

$$\left(\frac{\text{dln } a_{\text{A}}}{\text{dln } x_{\text{A}}}\right)_{T,P} = \left(\frac{\text{dln } a_{\text{B}}}{\text{dln } x_{\text{B}}}\right)_{T,P}$$

See also K.

Gibbs free energy function
See Free energy.

Gibbs–Helmholtz equation
The Gibbs–Helmholtz equation gives the variation of the free energy change, ΔG, of a chemical reaction occurring at constant pressure, with the temperature at which it is measured:

$$\left(\frac{\partial(\Delta G)}{\partial T}\right)_P = -\Delta S = \frac{\Delta G - \Delta H}{T}$$

i.e.

$$\Delta H - \Delta G = -T\left(\frac{\partial(\Delta G)}{\partial T}\right)_P \tag{G.1}$$

Hence, ΔH can be calculated from a knowledge of ΔG and the temperature coefficient of ΔG.

Alternative forms of equation (G.1) are

$$\left(\frac{\partial(\Delta G/T)}{\partial T}\right)_P = -\frac{\Delta H}{T^2} \text{ and } \left(\frac{\partial(\Delta G/T)}{\partial(1/T)}\right)_P = \Delta H$$

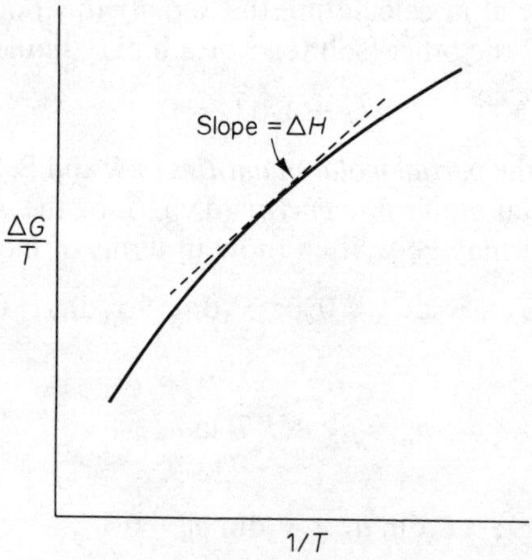

Figure G.3 Variation of $\Delta G/T$ with $1/T$ for a chemical reaction, the slope of the tangent to the curve at a given temperature is ΔH.

Thus, if $\Delta G/T$ is plotted against T^{-1} (figure G.3), the slope of the line is the *enthalpy* (q.v.) change at the given temperature.

For chemical reactions in reversible *galvanic cells*†

$$\Delta G = -nEF \text{ and } (\partial(\Delta G)/\partial T)_P = -nF(\partial E/\partial T)_P$$

which, on substitution into equation (G.1), gives

$$\Delta H + nFE = nFT(\partial E/\partial T)_P$$

The corresponding Gibbs–Helmholtz equation for a reaction occurring at constant volume is

$$\Delta U - \Delta A = -T\left(\frac{\partial(\Delta A)}{\partial T}\right)_V$$

H

Heat

Heat, q (dimensions: m l^2 t^{-2}; units: J), is a form of energy transferred between a system and its surroundings by thermal conduction, convection or electromagnetic radiation. It is as meaningless to say that a system contains heat as it is to say that it contains *work* (q.v.). Heat is an extensive *property* (q.v.) and always flows from a body of higher temperature to one of lower temperature. Heat transfer is an irreversible process. The heat, q, absorbed by the system from the surroundings during a change of state depends on the path (cf. *work*), i.e. both q and w are line integrals. Although $đw$ and $đq$ are not *exact differentials* (q.v.), nevertheless, for a given change of state, the algebraic sum given by

$$dU = đq - đw$$

is an exact differential.

See also First law of thermodynamics.

Heat capacity

The heat capacity, C, (dimensions: m l^2 t^{-2} deg^{-1}; units: J K^{-1}), of a system is the amount of heat required to raise its temperature by 1 K either at constant volume, C_V, or at constant pressure, C_p. The specific heat capacity is the heat capacity divided by the mass ($c_p = C_p/m$).

At constant volume, all the added heat goes to increase the *internal energy* (q.v.), U; thus

Heat capacity

$$C_V = (\partial U/\partial T)_V \tag{H.1}$$

Since the internal energy of an ideal gas is composed of contributions from translation ($\frac{3}{2}RT$), rotation ($\frac{3}{2}RT$ for non-linear and RT for linear molecules) and vibration ($[h\nu/(\exp(h\nu/kT)-1)]$ per vibrational mode), expressions can be readily obtained which allow the calculation of the translational, rotational and vibrational contributions to the heat capacity of the gas at constant volume. Thus the contribution from translation is $\frac{3}{2}R$, from rotation $\frac{3}{2}R$ (non-linear) or R (linear) and from vibration $Rx^2e^x/(e^x-1)^2$, where $x = h\nu/kT$. Thus, for the gas NO_2 at 298 K, the contributions are:

		$C/\text{J K}^{-1}\text{ mol}^{-1}$
Translation		12.47
Rotation		12.47
Vibration,	$\nu_1 = 2.25\times10^{13}\text{s}^{-1}, x = 3.63$	3.10
	$\nu_2 = 3.97\times10^{13}\text{s}^{-1}, x = 6.40$	0.59
	$\nu_3 = 4.84\times10^{13}\text{s}^{-1}, x = 7.80$	0.21
	C_V	28.84

At constant P, however, work has also to be done against the applied pressure; thus

$$C_p = (\partial H/\partial T)_P \tag{H.2}$$

For any system,

$$\begin{aligned} C_p - C_V &= \left(\frac{\partial H}{\partial T}\right)_P - \left(\frac{\partial U}{\partial T}\right)_V \\ &= \left[P + \left(\frac{\partial U}{\partial V}\right)_T\right]\left(\frac{\partial V}{\partial T}\right)_P \end{aligned} \tag{H.3}$$

since $U = f(T, V)$. The first term represents the contribution to C_p caused by the volume change against external pressure; the second term is the contribution from the energy required for the change in volume against the internal cohesive forces of the substance. For liquids and solids with large cohesive forces, the internal pressure $(\partial U/\partial V)_T$ is large, while for gases it is small compared with P, and so

$$C_p - C_V = P\left(\frac{\partial V}{\partial T}\right) = R$$

The difference is also expressed in the relationship

Table H.1. Molar heat capacity coefficients (valid between 300 and 1500 K)

	a/J K^{-1} mol^{-1}	$10^3\,b$/J K^{-2} mol^{-1}	$10^7\,c$/J K^{-3} mol^{-1}
NH_3	25.89	33.0	30.5
H_2	29.06	−0.837	2.01
N_2	27.29	5.23	−0.042
O_2	25.72	12.98	−38.61
CO	26.86	6.97	−8.20
CO_2	25.99	43.5	−148.3
H_2O	30.36	9.61	11.84

$$C_p - C_V = T\left(\frac{\partial U}{\partial V}\right)_T \left(\frac{\partial P}{\partial T}\right)_V = -T\left\{\frac{(\partial V/\partial T)_P^2}{(\partial V/\partial P)_T}\right\} = \frac{\alpha^2}{\beta} V\,T \qquad \text{(H.4)}$$

where α is the coefficient of expansion $[(\partial V/\partial T)_P/V]$ and β the coefficient of compressibility $[-(\partial V/\partial P)_T/V]$. For solids, the difference is small, although it becomes appreciable as α increases: e.g. $(C_p - C_V)$/J K^{-1} mol^{-1} for Li = 1.2; for C = 0.0; for S = 1.7; and for I = 3.8—values which, except for C, amount to about 10% of C_V. For liquids, the difference is greater than for the same substance in the solid state, since α is greater.

The heat capacity of a substance varies markedly with temperature; the variation is expressed in empirical equations of the type

$$C_p = a + bT + cT^2 + dT^3 \ldots \qquad \text{(H.5)}$$

where a, b, c, d ... are experimentally determined constants; typical values for these constants are listed (table H.1).

Numerical values of heat capacities of gases cannot be predicted by classical thermodynamics. With the aid of *statistical thermodynamics* (q.v.) and *partition functions* (q.v.), the heat capacity of an ideal gas can be calculated from spectroscopic data.

The heat capacity of a solid cannot be predicted by purely thermodynamic reasoning. The Debye equation, in its limiting form,

$$C_V = \frac{12\,\pi^4 R}{5}\,\frac{T^3}{\theta^3} = 1943.47\,\frac{T^3}{\theta^3}\ \text{J K}^{-1}\ \text{mol}^{-1}$$

(where θ, the characteristic temperature, is obtained from experimental heat capacity data at low temperatures) is useful in the extrapolation of measured heat capacities of solids down to the neighbourhood of 0 K. The molar heat capacity of solids is over 23 J K^{-1} when $T = \theta$, and approaches a value of 25 J K^{-1} asymptotically at higher temperatures.

Heat capacity

There is no adequate method of interpreting heat capacity data of liquids; it has generally been observed that the molar heat capacity of a liquid is near that of the solid.

Such empirical equations as (H.5) are useful in calculating (a) *enthalpy* (q.v.) changes at different temperatures using *Kirchhoff's equation* (q.v.), and (b) *entropy* (q.v.) changes for non-isothermal processes:

$$\Delta S = \sum \mathrm{d}S = \int \frac{\mathrm{d}H}{T} = \int C_p \,\mathrm{d}\ln T$$

Measurement of specific heat capacities of gases

Regnault's method at constant pressure A mass of gas, m, at known temperature, t_2, is passed at constant pressure, P, through a copper spiral in a water calorimeter; the increase in the temperature of the water is measured with all the usual precautions:

$$c_p = \frac{w}{m} \frac{\Delta t}{(t_2 - t_1)}$$

where w is the water equivalent of the calorimeter and contents, Δt is the increase in the temperature of the water, calorimeter, tubing etc., and t_1 is the temperature of the gas leaving the calorimeter.

Callendar and Barnes continuous flow method at constant pressure A stream of gas flowing at constant rate (m g s^{-1}) and constant pressure, P, through a tube is heated electrically by a wire in the tube. Of the electrical energy ($I \times V$) supplied, some goes to heat the gas through Δt and some, e, is lost to the surroundings; then

$$c_p = \frac{I \times V - e}{m\,\Delta t}$$

Allowance for loss to the surroundings can be made and, hence, c_p calculated.

Joly differential steam calorimeter at constant volume Two identical copper spheres, one filled with a known mass (m) of gas under pressure and the other evacuated, are suspended from the arms of a balance in a large, insulated calorimeter at temperature t. Steam is admitted and water condenses on both spheres; this is collected in trays attached to the spheres. More condenses on the sphere containing the gas, and the balance is again counterpoised. If m' is the additional mass of water condensed on the sphere containing the gas, then

$$c_V = \frac{m' \, l_e}{m(100 - t)}$$

Precautions must be taken to prevent loss of water from the spheres and dripping of water from the roof onto the spheres, and to ensure that the suspension wires are not coated with water. Correction has also to be made for the small change of volume.

Measurement of heat capacities of solids
The heat capacity of a solid is usually determined by the method of mixtures. A known weight of solid at a measured temperature is transferred to a weighed mass of liquid in a calorimeter, usually at a lower temperature, and the temperature change (corrected for losses by cooling curves) is determined. Normal calorimetric precautions must be taken. From the known water equivalent of the calorimeter and contents, the amount of heat gained (or lost) by the solid can be calculated and, hence, the heat capacity.

Measurement of heat capacities of liquids
The heat capacity of a liquid can be measured by the method of mixtures (as for solids), by use of a solid of known specific heat capacity, or by the continuous flow method (as for gases).

See also Equipartition of energy; Internal energy; and Ad, P (Vols. I, II, III).

Heat of combustion
See Enthalpy

Heat content
See Enthalpy.

Heat engine
A heat engine is a mechanism (or machine) which, working without frictional losses, obtains work from the passage of *heat* (q.v.) from a hotter to a colder body. When the process is carried out reversibly, the maximum amount of *work* (q.v.) is obtained from the machine.

Such engines consist of a source of heat maintained at a constant high temperature, T_2 (e.g. furnace or nuclear reactor), and a heat sink or cold reservoir maintained at a lower temperature, T_1 (e.g. atmosphere or cooling bath), which receives heat discarded by the engine as it operates. The heat engine operates by carrying a working substance through a cycle or sequence of operations which converts heat into work; at the end of each cycle the working substance has returned to its original states and a new cycle begins. The working substance usually approximates to an ideal gas, but, in principle, it can be any substance.

See also Carnot cycle.

Heat of formation

See Enthalpy

Helmholtz free energy

The Helmholtz free energy or maximum work function, A (dimensions: m l^2 t^{-2}; units: J mol^{-1}), is a *state function* (q.v.) defined by

$$A = U - TS$$

and is related to the Gibbs free energy, G (see *free energy*), by

$$G = A + PV$$

For an isothermal reversible process, the decrease in the work function is equal to the total *work* (q.v.) done by the process, i.e. $\mathrm{d}A = -đw$, or $\Delta A = -w$. If only PV work is involved, then $\mathrm{d}A = -P\mathrm{d}V$; thus, if the process is carried out at a constant volume (e.g. in a *bomb calorimeter*, q.v.) $\mathrm{d}A = \mathrm{d}V = 0$; hence, A is constant. For an isothermal process at constant volume, if $\Delta A < 0$, the process is spontaneous; if $\Delta A = 0$, the system is at equilibrium (A is a minimum); and if $\Delta A > 0$, the process as written is not spontaneous.

Since most chemical processes are carried out at constant pressure, G is a more useful function than A.

For a system doing no work other than $P\mathrm{d}V$ work,

$$\mathrm{d}A = -P\mathrm{d}V - S\mathrm{d}T$$

Hence,

$$\left(\frac{\partial A}{\partial V}\right)_T = -P \text{ and } \left(\frac{\partial A}{\partial T}\right)_V = -S$$

or

$$\left(\frac{\partial(\Delta A)}{\partial V}\right)_T = -\Delta P \text{ and } \left(\frac{\partial(\Delta A)}{\partial T}\right)_V = -\Delta S$$

Thus,

$$\Delta A = \Delta U - T\Delta S = \Delta U + T\left(\frac{\partial(\Delta A)}{\partial T}\right)_V$$

See also Dic, K, M & P.

Henderson equation

The Henderson equation is used to calculate the pH of a *buffer solution* (q.v.), which consists of a mixture of a weak acid (HA) and its conjugate base in the form of a fully dissociated salt (Na^+A^-). The equation

$$\mathrm{pH} = \mathrm{p}K_a + \log \frac{c(\mathrm{A}^-)}{c(\mathrm{HA})}$$

is obtained from the definition of the acid *dissociation constant* (q.v.).

The equation shows that the pH is determined by the acid dissociation constant and the ratio of the concentrations of the salt and free acid, and predicts that the pH is independent of the actual concentrations. The pH of a buffer solution varies with the concentration; the discrepancy between the experimental observations and the theoretical prediction is due to the neglect of *activity coefficients* (q.v.) in deriving the Henderson equation.

Henry's law

At a fixed temperature, the weight of gas (w_B) dissolved in a given volume of solvent (V_A) is proportional to the partial pressure, i.e. $w_B/V_A = k\ p_B$, which, in terms of mole fractions, may be written

$$x_B = k' p_B$$

or, more generally,

$$p_B = k_B x_B$$

where k_B is the proportionality constant.

For *ideal solutions* (q.v.), *Raoult's law* (q.v.) and Henry's law become identical, i.e. $k_B = p_B^{\ominus}$ and $k_A = p_B^{\ominus}$.

See also Real solutions.

Hess's law of constant heat summation

Hess's law of constant heat summation states that the total heat effect (i.e. change in *enthalpy*, q.v.) in any series of physical and chemical changes is independent of the path by which the system goes from the initial to the final state.

Although enunciated in 1840, it is now recognised as a corollary of the *first law of thermodynamics* (q.v.). Since U and V are *state functions* (q.v.) and P is constant, $H\ (= U + PV)$ must also be a state function.

Hess's law is useful in calculating heats of reaction from known enthalpies of formation and combustion; many of these reactions proceed too slowly to be measured directly, and in some there are undesired side reactions. Consider the application of the law to the calculation of the enthalpy of formation of methane at 298 K according to the scheme

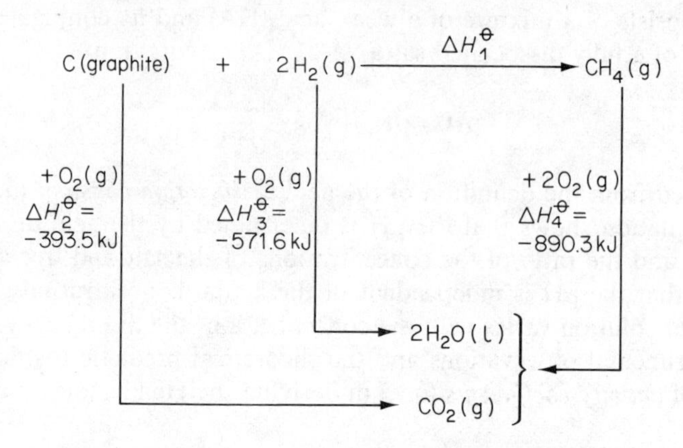

Since the reactants C, $2H_2(g)$ and $2O_2(g)$ and the products $2H_2O(l)$ and $CO_2(g)$ are identical, then, according to Hess's law,

$$\Delta H_1^\ominus + \Delta H_4^\ominus = \Delta H_2^\ominus + \Delta H_3^\ominus$$

and, hence,

$$\Delta H_1^\ominus = -393.5 - 571.6 - (-890.3) = -74.8 \text{ kJ}$$

It is customary to arrange the equations for direct addition and cancellation of terms; thus, for the enthalpy of formation of acetylene at 298 K,

$$\text{C(graphite)} + O_2 \longrightarrow CO_2(g) \qquad \Delta H^\ominus = -393.5 \text{ kJ} \quad (H.6)$$

$$H_2(g) + \tfrac{1}{2}O_2(g) \longrightarrow H_2O(l) \qquad \Delta H^\ominus = -285.8 \text{ kJ} \quad (H.7)$$

$$C_2H_2(g) + \tfrac{5}{2}O_2(g) \longrightarrow 2CO_2(g) + H_2O(l) \qquad \Delta H^\ominus = -1299.5 \text{ kJ} \quad (H.8)$$

(H.6) × 2 + (H.7) – (H.8) gives

$$2\text{C(graphite)} + H_2(g) \longrightarrow C_2H_2(g)$$

and

$$\Delta H_f^\ominus(C_2H_2, g) = 2 \times (-393.5) - 285.8 - (-1299.5) = 226.7 \text{ kJ}$$

When the enthalpies of combustion or formation of reactants and products are known, the heat of reaction can be calculated from

$$\begin{aligned}\Delta H^{\ominus} &= \sum \Delta H_c^{\ominus}(\text{reactants}) - \sum \Delta H_c^{\ominus}(\text{products}) \\ &= \sum \Delta H_f^{\ominus}(\text{products}) - \sum \Delta H_f^{\ominus}(\text{reactants})\end{aligned}$$

See also Born–Haber cycle.

High-energy bonds

In biochemical language, a high-energy bond is one which liberates an unusually large amount of *free energy* (q.v.) when it is broken by hydrolysis. Adenosine triphosphate, ATP, consisting of an adenine base, a five-carbon sugar (ribose) and three phosphate groups, is an effective storage device in the living organism for chemical energy, since two of the phosphate groups have an unusually high free energy of formation. This stored energy is released for use when required, e.g. in muscular contraction. The net result of one complete round of the Kreb's cycle is the combustion of 1 mole of pyruvic acid; if this is achieved directly, the efficiency of the process is only 5.5%; this is increased to 48% when the process is carried out stepwise in the cycle. ATP is one of the components of this free energy transfer process.

This compound is sometimes written A—P~P~P, where the wavy lines represent 'high-energy' bonds, which on hydrolysis liberate large amounts of energy, thus:

$$\text{A—P}\sim\text{P}\sim\text{P} + H_2O \longrightarrow \underset{\text{(ADP)}}{\text{A—P}\sim\text{P}} + HPO_4^{2-} + H^+ \quad \Delta G' = -29.3\ \text{kJ (high energy)}$$

$$\text{A—P}\sim\text{P} + H_2O \longrightarrow \underset{\text{(AMP)}}{\text{A—P}} + HPO_4^{2-} + H^+ \quad \Delta G' = -29.3\ \text{kJ (high energy)}$$

$$\text{A—P} + H_2O \longrightarrow \text{A} + HPO_4^{2-} + H^+ \quad \Delta G' = -12.5\ \text{kJ (normal)}$$

$\Delta G'$ means that the standard free energy change, $\Delta G^{\ominus}$, is calculated at the physiological pH of 7.

This large free energy of hydrolysis of such compounds is attributed to the increase in resonance stabilisation and decrease in electrostatic repulsion. In biological systems, this energy is not wasted in fruitless hydrolysis reactions but is either: (1) conserved by transfer with little loss of energy, as in the formation of creatine phosphate in muscle cells:

$$\text{ATP} + \text{creatine} \xrightarrow{\text{creatine phosphokinase}} \text{ADP} + \text{creatine}\sim\text{P} + H^+ \quad \Delta G' = +11.7\ \text{kJ}$$

although $\Delta G'$ is unfavourable, the accumulation of ATP in resting muscle causes the reaction to go forward, thus storing high-energy phosphate until required; or (2) utilised in the synthesis of necessary metabolites, as in the synthesis of glycerol-1-phosphate, which cannot be accomplished by the esterification of glycerol with inorganic phosphate. The reaction, catalysed by α-glycerokinase, can be imagined as proceeding:

$$\begin{array}{lll} \text{ATP} + \text{H}_2\text{O} & \longrightarrow \text{ADP} + \text{H}^+ + \text{P} & \Delta G' = -29.3\ \text{kJ} \\ \text{Glycerol} + \text{P} & \longrightarrow \text{Glycerol-1-P} + \text{H}_2\text{O} & \Delta G' = +\ 9.6\ \text{kJ} \\ \hline \text{ATP} + \text{Glycerol} & \longrightarrow \text{ADP} + \text{Glycerol-1-P} + \text{H}^+ & \Delta G' = -\ 19.7\ \text{kJ} \end{array}$$

in which the high negative free energy of hydrolysis of ATP is the driving force of the reaction.

See also Dic, W & W.

Hydrolysis of salts
See Acids and bases.

I

Ideal solutions
Ideal solutions are characterised by (1) vapour pressure–concentration curves which conform to *Raoult's law* (q.v.), (2) a zero heat of mixing and (3) a solution volume which is the sum of the component volumes.

If 1 mole of an ideal solution is formed from x_A mole of A and x_B mole of B, the net change in free energy from unmixed components to mixed components is given by

$$\Delta G(\text{mixing}) = x_A \Delta G_A + x_B \Delta G_B$$

but, for the individual components,

$$\Delta G_A = \mu_A(\text{solution}) - \mu_A^\ominus = RT \ln x_A$$

whence

$$\Delta G(\text{mixing}) = x_A RT \ln x_A + x_B RT \ln x_B$$

$\Delta G(\text{mixing}) < 0$ and, for a given mole fraction, is independent of the nature of the components. Since $\Delta H(\text{mixing}) = 0$, it follows that there is an increase of entropy, on mixing, given by

$$\Delta S(\text{mixing}) = -x_A R \ln x_A - x_B R \ln x_B$$

See also Entropy; Real solutions.

Immiscible liquids

Immiscible liquids are pairs of liquids for which the mutual solubilities are so small as to be insignificant. Two layers exist and each liquid exerts its own v.p. independently of the presence of the other liquid; the total v.p. is the sum of those of thc components (figure I.1). From the *phase rule* (q.v.), since $c = 2$, $p = 3$ and, hence, $f = 1$, i.e. at a given temperaturc the total v.p. is constant for mixtures of all compositions.

Assuming ideal behaviour of the vapour phase,

$$\frac{p_A}{p_B} = \frac{n_A}{n_B} = \frac{w_A/M_r(A)}{w_B/M_r(B)}$$

where n_A and n_B are the number of moles of A and B, respectively, in the vapour phase; hence,

$$\frac{\text{Weight of A in vapour}}{\text{Weight of B in vapour}} = \frac{w_A}{w_B} = \frac{p_A M_r(A)}{p_B M_r(B)}$$

If a mixture is heated until the total pressure is atmospheric, then distillation occurs at a temperature below the b.p. of either pure A or pure B, while the two layers still exist. The relative weights of A and B which distil over are given by the equation. When water is one of the liquids, the process is known as *steam distillation* (q.v.).

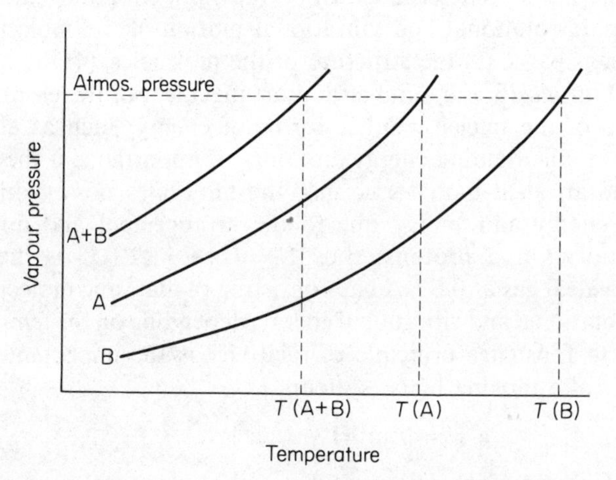

Figure I.1 Vapour pressure–temperature curves for pure components A and B and a mixture of A and B.

Incongruent melting point
A compound is said to have an incongruent or meritectic *melting point* (q.v.) when the compound is not very stable even in the solid state and decomposes below its melting point. The solid cannot, therefore, exist in equilibrium with liquid of the same composition and is said to be incongruent. In such *two-component condensed systems* (q.v.), the solid establishes equilibrium with a second solid phase and a liquid of different composition.

Inexact differential
An inexact differential, e.g. $đq$ or $đw$, cannot be obtained by differentiating a *state function* (q.v.) and cannot be integrated to give q or w, respectively.
See also Exact differential.

Intensity factor
See Property.

Intensive property
See Property.

Internal energy
The internal energy, U (dimensions: $\mathrm{m\,l^2\,t^{-2}}$; units: $\mathrm{J\,mol^{-1}}$), of every system has a value which is a function of the chemical nature of the system, the temperature and often the pressure and volume of the system. For a given system of molecules, U is determined by (1) the number of molecules, (2) the kinetic (thermal), rotational and vibrational motion of the molecules and their component parts, (3) the structure of the molecules, (4) the nature of the individual atoms, (5) the number and arrangement of the electrons and (6) the nature of the nucleus. Other forms of energy, such as electronic excitation or atomic binding energy, are only of importance if they can be tapped, e.g. in an ideal monatomic gas, the molecules possess kinetic or translational energy and energy due to the arrangement and number of electrons, neutrons and protons; thus $U = U_0 + \frac{3}{2}RT$ (U_0 is the energy content of the ideal gas at 0 K). A gas consisting of diatomic molecules may also possess rotational and vibrational energy, depending on the temperature.

Embodied in Einstein's principle of relativity is the important relation for the energy of a moving body (system):

$$U = m_0c^2/(1 - v^2/c^2)^{1/2}$$

where m_0 is the mass of the system at rest, v the velocity of the system and c the velocity of light; when $c \gg v$,

$$U = m_0c^2 + \tfrac{1}{2}m_0v^2$$

In this equation the last term is the kinetic energy of the moving body and the first term the energy of the body at rest. Theoretically, this provides a fundamental basis for the assignment of absolute values of U. Such an assignment adds large and unwieldy constants to the numbers to be added and subtracted in thermodynamic processes. Since thermodynamics is concerned only with values of ΔU in specified physical or chemical processes, the zero from which the energy changes are measured is quite arbitrary.

Einstein's theory relates change of mass with change of energy; thus

$$\Delta U = c^2\,\Delta m_0 = 8.98 \times 10^{13}\,\Delta m_0\ \mathrm{J}$$

i.e. a gain in mass of 1 g corresponds to a gain in U of 8.98×10^{13} J.

U is a *state function* (q.v.) and so ΔU is independent of the path taken, i.e.

$$\Delta U = U\,(\text{final}) - U\,(\text{initial})$$

and, for a closed cycle,

$$\oint \mathrm{d}U = 0$$

For *isothermal processes* (q.v.), in which there is a change of volume at constant pressure, $w = P\Delta V$, and the application of the *first law of thermodynamics* (q.v.) gives

$$q_P = \Delta U + P\Delta V = \Delta H$$

At constant volume, $q_V = \Delta U$.

Since $U = f(T,V)$, it follows that when n mole of an ideal gas expand isothermally, $\Delta U = 0$, and, hence,

$$q = w = nRT \ln V_2/V_1 = nRT \ln P_1/P_2$$

For a chemical reaction, ΔU varies with temperature according to *Kirchhoff's equation* (q.v.):

$$\Delta U(T_2) = \Delta U(T_1) + \Delta C_V(T_2 - T_1)$$

U and ΔU are related to other thermodynamic quantities by the equations

$$H = U + PV\,;\ U = A + TS$$

$$C_V = \left(\frac{\partial U}{\partial T}\right)_V\,;\ \left(\frac{\partial \ln K_c}{\partial T}\right)_V = \frac{\Delta U^\ominus}{RT^2}$$

Internal energy

$$U = -T^2\left(\frac{\partial(A/T)}{\partial T}\right)_{V,N}$$

$$\left(\frac{\partial U}{\partial V}\right)_S = -P\,; \left(\frac{\partial U}{\partial S}\right)_V = T$$

$$\left(\frac{\partial U}{\partial T}\right)_P = \left(\frac{\partial S}{\partial P}\right)_T$$

In statistical thermodynamics,

$$U = N_A kT^2\left(\frac{\partial \ln Q}{\partial T}\right)_V$$

Calculation and measurement of ΔU
ΔU may be calculated in the following ways.

(1) Energy change on heating an ideal gas: $\Delta U = C_V(T_2 - T_1)$.

(2) For a spontaneous change in which at least one reactant is gaseous, the *bomb calorimeter* (q.v.) is used.

(3) For reactions which cannot be carried out directly, the over-all value of ΔU can be calculated from the values of ΔU for the related reactions; this follows since U is a state function (cf. *Hess's law of constant heat summation*).

(4) From a knowledge of the variation of K_c with temperature (see *van't Hoff isochore*).

See also Dnb, K, War.

Invariant system
An invariant system is one in which there are sufficient algebraic equations to permit the complete definition of all the variables. According to the *phase rule* (q.v.), when $f = 0$, there are no degrees of freedom and the system at equilibrium is completely defined or invariant.

For a *one-component system* (q.v.), $f = 0$ when $p = 3$, i.e. the system is completely defined at the point where all three phases co-exist in equilibrium—the *triple point* (q.v.).

In *two-component condensed systems* (q.v.) at constant pressure, the *eutectic point* (q.v.), at which three phases are in equilibrium, is an invariant point. Similarly, in *three-component systems* (q.v.), at constant pressure, the ternary eutectic point is an invariant point.

Inversion temperature
See Joule–Thomson coefficient.

Ion
See Enthalpy; Entropy; Free energy; and D & J.

Ionic product of water
Pure water has a small electrical *conductivity*† due to self-ionisation:

$$H_2O + H_2O \rightleftharpoons H_3O^+ + OH^-$$

This process of proton transfer is known as *autoprotolysis* (q.v.). For this equilibrium, since $a(H_2O) = 1$,

$$K_w = a(H_3O^+)\, a(OH^-), \quad pK_w = -\log K_w$$

The ionic product principle must be valid in any aqueous solution, independent of its pH. For all acids and conjugate bases in aqueous solution, the product of the acidic and basic *dissociation constants* (q.v.) equals K_w.
The value of K_w has been determined from accurate conductivity measurements on conductivity water. It may also be determined from e.m.f. studies on *cells*† without *liquid junctions*†, e.g.

$$\ominus \text{ Pt, } H_2(\text{g, } 101\,325 \text{ N m}^{-2}) \left| \begin{matrix} \text{KOH, KCl} \\ m_1 \quad m_2 \end{matrix} \right| \text{AgCl, Ag } \oplus$$

for which,

$$\frac{F[E - E^{\ominus}(\text{Ag, AgCl, Cl}^-)]}{2.303\, RT} + \log\frac{m_2}{m_1} = -\log K_w - \log\frac{\gamma(\text{Cl}^-)}{\gamma(\text{OH}^-)}$$

The value of E for the cell is measured for known values of m_1 and m_2; the graph of the left-hand side of the equation against $I^{1/2}$ is linear and of intercept $-\log K_w$.

Table I.1

$T/°C$	$10^{14}\, K_w/\text{mol}^2\,\text{dm}^{-6}$
0	0.116
10	0.281
18	0.590
25	1.008
40	2.919
50	5.660
60	9.614

Ionic product of water

The ionic product of water varies with temperature (table I.1).

Since at 25 °C, $K_w \approx 10^{-14}$ mol^2 dm^{-6}, the concentrations (activities) of the H_3O^+ and OH^- are equal to 10^{-7} mol dm^{-3}, i.e. the pH of pure water is 7.0.

Ionic strength

Ionic strength, I (units: mol dm^{-3}, mol m^{-3}, mol kg^{-1}), is a characteristic of an electrolyte solution defined by

$$I = \tfrac{1}{2}\sum_i m_i z_i^2$$

where m_i represents the molality of the ith ion and z_i its charge; the summation extends over all ions in solution. In terms of volume concentration c_i/mol dm^{-3}, the ionic strength I/mol dm^{-3} is given by

$$I = \tfrac{1}{2}\sum_i c_i z_i^2$$

Thus, for a solution of Na_2SO_4 of molality represented by m,

$$I/\text{mol kg}^{-1} = \tfrac{1}{2}[m(Na^+)z^2(Na^+) + m(SO_4^{2-})z^2(SO_4^{2-})]$$
$$= \tfrac{1}{2}(2m + 4m) = 3m$$

Using this quantity, it is possible to compare the effects of salts of different valence type. The relation between I and molality for salts of different valence type is tabulated in table I.2.

Table I.2

Valence type	Example	I
1:1	NaCl	m
1:2	Na_2SO_4	$3m$
2:1	$CaCl_2$	$3m$
1:3	Na_3PO_4	$6m$
3:1	$AlCl_3$	$6m$
2:2	$MgSO_4$	$4m$

Ionisation constant

See Dissociation constant.

Ionisation energy

The ionisation energy, I of a gaseous atom or ion is the increase in internal energy (ΔU_0) at 0 K which accompanies the ionisation process and the formation of a gaseous cation. It is a measure of the energy which is required to remove successive electrons in the stepwise process:

$$\begin{aligned} M(g) &\longrightarrow M^+(g) + e \\ M^+(g) &\longrightarrow M^{2+}(g) + e \\ M^{2+}(g) &\longrightarrow M^{3+}(g) + e \end{aligned}$$

Successive ionisation energies (in kJ mol^{-1}) or ionisation potentials (in eV) for the electrons of a given atom always increase, because after the first, each additional electron is removed from a gaseous cation carrying an increasing positive charge. The first ionisation energy decreases monotonically down groups 1, 2, 6, 7 and 0 and shows a general upward trend from left to right across each period.

Ionisation energies are encountered in enthalpy cycles which refer to constant pressure systems at 298 K. The enthalpy of ionisation is related to the ionisation energy at 0 K by the application of *Kirchhoff's equations* (q.v.):

$$\Delta H(298\ \mathrm{K}) = \Delta U_0 + \int_0^{298} [C_p(\mathrm{M}^+) + C_p(\mathrm{e}) - C_p(\mathrm{M})]\,\mathrm{d}T$$

Assuming that M(g), M^+(g) and e(g) can be regarded as ideal monatomic gases, then their heat capacities may be taken as 0 at 0 K and $5R/2$ at any other temperature. Thus

$$\Delta H(298\ \mathrm{K}) = \Delta U_0 + \tfrac{5}{2}RT$$

At 298 K the correction amounts to about 6.2 kJ mol^{-1}.

In such cycles both an *electron affinity* (q.v.) and an ionisation energy normally contribute, and if the number of electrons involved in each term is the same, then the correction terms cancel out and the ΔU_0 values can be used.

See also table A.IV (p. 253); and Da.

Irreversible process

An irreversible process is one in which the reversal of the process cannot be achieved by simply changing the parameters by an infinitesimal amount (cf. *reversible process*). Irreversibility does not imply that it is impossible to force a process in the reverse direction. All naturally occurring or spontaneous processes, even the very slow ones, are irreversible, e.g. the flow of heat from a hotter to a colder body, the solution of salt in water, the expansion of a gas into a vacuum. *Work* (q.v.) obtained in an irreversible process is

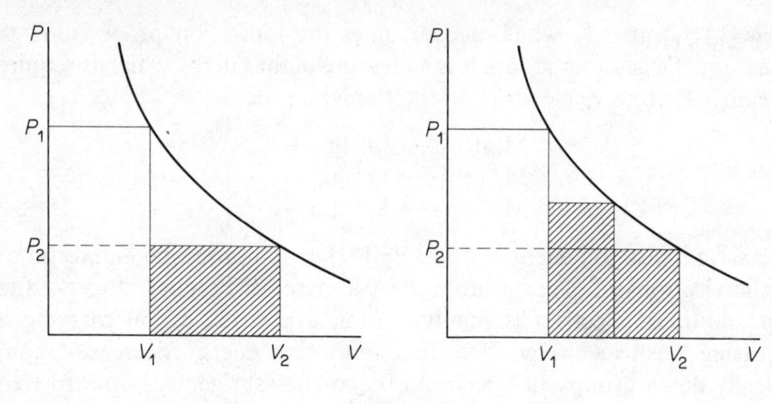

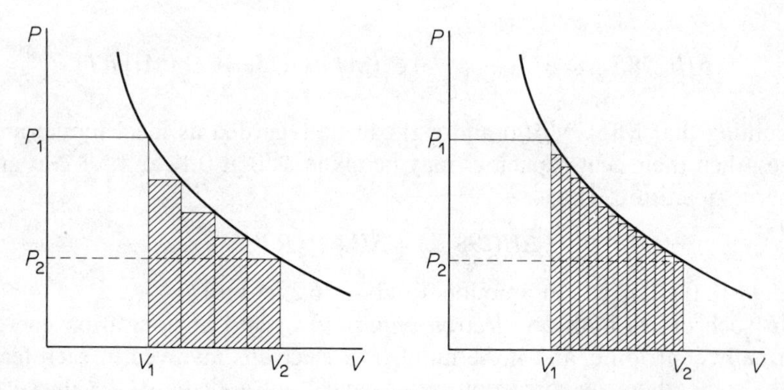

Figure I.2 Isothermal expansion of a gas, showing a few of the many possible irreversible paths which may be taken.

always less than the maximum obtained when the process is carried out reversibly, since there is no unique and continuous relationship between the parameters. For example, consider the irreversible isothermal expansion of a gas from P_1V_1 to P_2V_2; this can be achieved by a large number of paths (figure I.2) and the work ($P\Delta V$) obtained from each differs. As the number of steps in the process increases, the work done (shaded area) increases, and in the limit becomes equal to that obtained when the process is carried out reversibly.

For any change involving an ideal gas, irrespective of whether the process is reversible or irreversible,

$$\mathrm{d}U = C_V\,\mathrm{d}T \text{ and } \mathrm{d}H = C_p\,\mathrm{d}T$$

See also Was.

Isenthalpic process
An isenthalpic process is a process carried out under such conditions that there is no change in *enthalpy* (q.v.), i.e. $\mathrm{d}H = 0$. The best example of such a process is the throttled expansion of a gas under adiabatic conditions (see *adiabatic process*), when there is a small change in the temperature of the gas. The *Joule–Thomson coefficient* (q.v.) represents the change in temperature per unit change in pressure, i.e.

$$\mu_{\mathrm{JT}} = (\partial T/\partial P)_H$$

Isentropic process
See Adiabatic process.

Isobaric process
An isobaric process is a process in which the pressure remains constant, i.e. $\mathrm{d}P = 0$. For a reversible isobaric process (see *reversible process*) involving an ideal gas,

$$\mathrm{d}U = C_V\,\mathrm{d}T;\ \mathrm{d}H = C_p\,\mathrm{d}T;\ \text{đ}w = P\,\mathrm{d}V$$

and, for a finite change,

$$q = \Delta H = \Delta U + P\Delta V = \int C_p\,\mathrm{d}T$$

See also Isochoric process (figure I.3).

Isochoric process
An isochoric process is a process in which the volume remains constant, i.e. $\mathrm{d}V = 0$. For a reversible isochoric process (see *reversible process*) involving an ideal gas (figure I.3),

$$\mathrm{d}U = C_V\,\mathrm{d}T,\ \mathrm{d}H = C_p\,\mathrm{d}T,\ \text{đ}w = 0$$

and, for a finite change,

$$q = \Delta U = \int C_V\,\mathrm{d}T = \int T\,\mathrm{d}S$$

Isochoric process

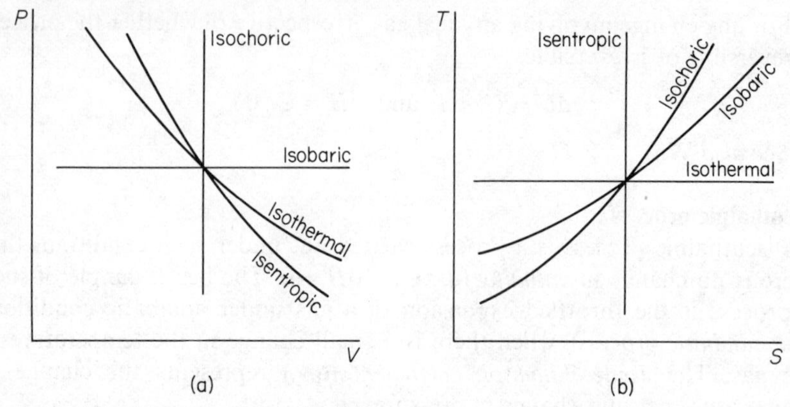

Figure I.3 P–V (a) and T–S (b) plots for isobaric, isochoric, isentropic and isothermal processes.

See also Isobaric process; Isothermal process.

Isoteniscope
See Vapour pressure.

Isothermal process
An isothermal process is one in which the temperature remains constant, i.e. $dT = 0$. To maintain isothermal conditions, reactions are carried out in constant temperature baths or thermostats.

The isothermal expansion of n mole of an ideal gas may be (1) reversible (see *reversible process*), (2) free, i.e. expansion into a vacuum, or (3) actual, in which some work is obtained, since the expansion is carried out at a finite speed. In (3) the pressure adjacent to the piston wall is slightly less than it is in the interior, and so the *work* (q.v.) obtained is less than in a reversible expansion but > 0. The values of the thermodynamic change, tabulated for the three types of expansion in table I.3, emphasise the dif-

Table I.3. Thermodynamic changes in isothermal expansions of an ideal gas

Reversible	Free	Actual
$T_1 = T_2$	$T_1 = T_2$	$T_1 = T_2$
$w = nRT \ln P_1/P_2$	$w = 0$	$0 < w < nRT \ln P_1/P_2$
$\Delta U = 0$	$\Delta U = 0$	$\Delta U = 0$
$q = nRT \ln P_1/P_2$	$q = 0$	$0 < q < nRT \ln P_1/P_2$
$\Delta H = 0$	$\Delta H = 0$	$\Delta H = 0$

ference between an exact (U or H) and an inexact (q or w) thermodynamic function.

See also Isobaric process; Isochoric process (figure I.3).

J

Joule

The joule, J (dimensions: $m\,l^2\,t^{-2}$; units: $kg\,m^2\,s^{-2}$), is the SI unit of energy, defined as the *work* (q.v.) done when the point of application of a force of 1 newton is displaced through a distance of 1 metre in the direction of the force, i.e. $J = N\,m$. (1 calorie = 4.184 J.)

Joule–Thomson coefficient

When a gas passes through a throttle from a high- to a low-pressure region under adiabatic conditions ($q = 0$), there is a small but measurable change in the temperature of the gas.

$\mu_{JT} = (\partial T/\partial P)_H$ or, for a finite change, $\mu_{JT} = (\Delta T/\Delta P)_H$, where μ_{JT} (units: $K\,m^2\,N^{-1}$) represents the change in temperature for a decrease in pressure of $1\,N\,m^{-2}$, for the process occurring at constant H. The temperature falls as the gas passes through a porous plug, because the attraction between the molecules retards their escape, thereby diminishing the kinetic energy and, hence, the temperature. For an ideal gas, μ_{JT} is zero. For hydrogen above 193 K, and helium above 53 K, the effect of internal attraction is negligible compared with that of molecular size; this then acts as repulsion and gives a negative Joule–Thomson effect. μ_{JT} depends on the temperature and pressure—in general, decreasing with increasing pressure—and becomes negative at high pressures. At low pressure, μ_{JT} is negative at low temperatures, increasing through 0 to a maximum positive value at an intermediate temperature, finally decreasing through 0 to a negative value at high temperatures. At a definite pressure, the values of temperature at which μ_{JT} is zero are known as the upper and lower inversion temperatures.

From the *thermodynamic equations of state* (q.v.),

$$\mu_{JT} = (\partial T/\partial P)_H = -\frac{(\partial H/\partial P)_T}{(\partial H/\partial T)_P} = \frac{T(\partial V/\partial T)_P - V}{C_p}$$

For a gas obeying *van der Waals' equation* (q.v.),

$$\mu_{JT} = \frac{1}{C_p}\left(\frac{2a}{RT} - b\right)$$

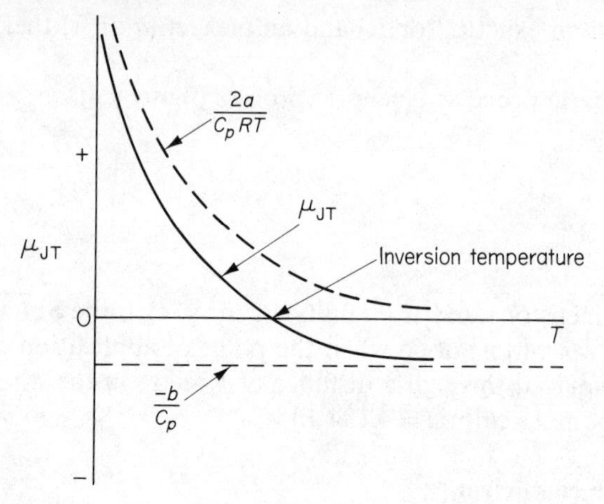

Figure J.1 The two components of μ_{JT} for a van der Waals gas.

If $b < 2a/RT$, then μ_{JT} is positive; if $b = 2a/RT$, $\mu_{JT} = 0$; and if $b > 2a/RT$, μ_{JT} is negative (figure J.1).

See also Dic.

K

Kelvin

The kelvin, K, the unit of thermodynamic temperature, is the fraction 1/273.16 of the thermodynamic temperature of the *triple point* (q.v.) of water. K is used both for thermodynamic temperature and for the thermodynamic temperature interval, e,g.

$$T = 305.53 \text{ K}, T(\text{ice}) = 273.15 \text{ K}, T - T(\text{ice}) = 32.38 \text{ K}$$

See also McG.

Kirchhoff's equations

If $\Delta H(T_1)$ is known for a reaction at T_1, then $\Delta H(T_2)$ can be calculated, provided the *heat capacity* (q.v.) values of reactants and products are known. Schematically:

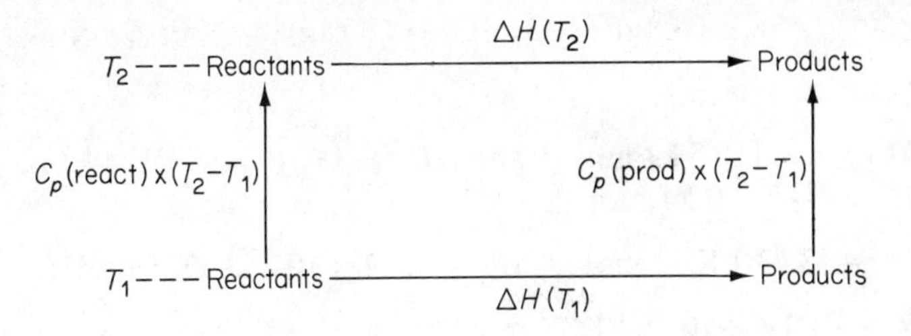

Assuming C_p(react) and C_p(prod), i.e. the sums of the heat capacities of reactants and products, respectively, are constant over the range T_1 to T_2, then, from *Hess's law of constant heat summation* (q.v.),

$$\Delta H(T_1) + C_p(\text{prod}) \times (T_2 - T_1) = \Delta H(T_2) + C_p(\text{react}) \times (T_2 - T_1)$$

or

$$\begin{aligned}\Delta H(T_2) - \Delta H(T_1) &= [C_p(\text{prod}) - C_p(\text{react})] \times (T_2 - T_1) \\ &= \Delta C_p (T_2 - T_1)\end{aligned}$$

Alternatively,

$$(\partial(\Delta H)/\partial T)_P = \Delta C_p$$

When C_p is not constant over the temperature range, the heat capacity for each reactant and product must be expressed in terms of an empirical equation of the form:

$$C_p = a + bT + cT^2 \ldots \tag{H.5}$$

whence,

$$\mathrm{d}(\Delta H)/\mathrm{d}T = \Delta a + \Delta bT + \Delta cT^2 \ldots$$

where

$$\Delta a = \sum a(\text{prod}) - \sum a(\text{react}), \text{ etc.}$$

On integration between limits, this gives

$$\Delta H(T_2) - \Delta H(T_1) = \Delta a(T_2 - T_1) + \tfrac{1}{2}\Delta b(T_2^2 - T_1^2) + \tfrac{1}{3}\Delta c(T_2^3 - T_1^3) \ldots$$

When $T_1 = 0$ K, this equation becomes

$$\Delta H(T) = \Delta H_0 + \Delta aT + \tfrac{1}{2}\Delta bT^2 + \tfrac{1}{3}\Delta cT^3 \ldots$$

Thus, for the reaction,

Kirchhoff's equations

$$CO(g) + \tfrac{1}{2}O_2(g) \longrightarrow CO_2(g)$$

at 298 K,

$\Delta H_f^\ominus(CO, g) = -110.5\ \mathrm{kJ\ mol^{-1}}$ and $\Delta H_f^\ominus(CO_2, g) = -393.5\ \mathrm{kJ\ mol^{-1}}$ whence $\Delta H^\ominus = -283.0\ \mathrm{kJ\ mol^{-1}}$.

$$\Delta a = -13.73\ \mathrm{J\ K^{-1}\ mol^{-1}};\ \Delta b = 30.04 \times 10^{-3}\ \mathrm{J\ K^{-2}\ mol^{-1}}$$

$$\Delta c = -120.8 \times 10^{-7}\ \mathrm{J\ K^{-3}\ mol^{-1}}$$

Hence,

$$\Delta H^\ominus(T) = -283\,000 + \int_{298}^{T} (-13.73 + 30.04 \times 10^{-3}\,T - 120.8 \times 10^{-7}\,T^2)\,\mathrm{d}T$$

This equation can be integrated to obtain $\Delta H^\ominus(T)$ at any temperature over which the heat capacity equations are valid.

The analogous equation for the calculation of ΔU at different temperatures is

$$\Delta U(T_2) - \Delta U(T_1) = \Delta C_V\,(T_2 - T_1)$$

See also Dic, Ro.

L

Landsberger's method

In this method for the determination of the elevation of the *boiling point* (q.v.), the solution in the apparatus is heated by vapour passing in from the boiling solvent (figure L.1). Superheating is theoretically impossible, since, when the b.p. is attained, the liquid is in equilibrium with the vapour at atmospheric pressure and no more will condense except to supply heat lost by radiation.

The b.p. of the solvent is determined at atmospheric pressure and then the b.p. of the solution after the addition of a known weight of solute, w_B. Bubbling is now stopped and the volume, V_A/m^3, of the dilute solution is recorded.

The equation relating the elevation of the boiling point and concentration:

$$\Delta T_e = k_e m_B$$

can be rewritten

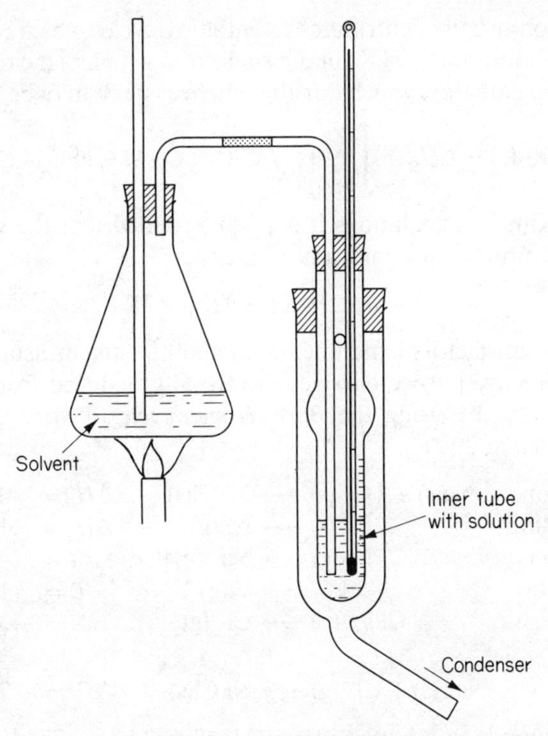

Figure L.1 Landsberger's apparatus for the measurement of the elevation of the boiling point of a solvent.

$$\Delta T_e = \frac{k_e \, w_B}{M_B \, V_A \, \rho_A} = \frac{k' w_B}{M_B \, V_A}$$

where the constant $k' = k_e/\rho_A$ is usually determined by calibration with a solute of known relative molecular mass. Thus it is possible to determine M_B; the result is, however, only of limited accuracy because of the inaccuracy involved in measuring the volume of the solution, which is further assumed to be the volume of the solvent.

Lattice energy

The lattice energy of a crystal is the energy liberated when 1 mole of the substance is formed from its constituents in the gaseous phase at 0 K, e.g. in the process

$$M^+(g) + X^-(g) \longrightarrow MX(s)$$

This reaction usually contributes to enthalpy cycles which refer to constant pressure conditions at 298 K; under such circumstances the relevant quantity is the lattice enthalpy, which, for the above system, may be calculated:

$$\Delta H(298\ \text{K}) = \Delta U_0 + \int_0^{298} [C_p(\text{MX}) - C_p(\text{M}^+, \text{g}) - C_p(\text{X}^-, \text{g})]\, \text{d}T$$

For approximate calculations it is often assumed that the variation of ΔU with temperature is insignificant and, hence,

$$\Delta H(298\ \text{K}) = \Delta U_0 - 2RT$$

There is no satisfactory experimental method for the measurement of lattice energies directly. Lattice energies are usually deduced from enthalpies of formation, etc., by using the *Born–Haber cycle* (q.v.); e.g., for NaCl (all values quoted at 298 K),

Formation	$\text{Na(s)} + \frac{1}{2}\text{Cl}_2\text{(s)} \longrightarrow \text{NaCl(s)}$	$\Delta H_\text{f} = -414\ \text{kJ mol}^{-1}$
Sublimation	$\text{Na(s)} \longrightarrow \text{Na(g)}$	$\Delta H_\text{a} = 109\ \text{kJ mol}^{-1}$
Ionisation	$\text{Na(g)} \longrightarrow \text{Na}^+\text{(g)} + \text{e}$	$I = 490\ \text{kJ mol}^{-1}$
Dissociation	$\frac{1}{2}\text{Cl}_2 \longrightarrow \text{Cl(g)}$	$\frac{1}{2}D = 113\ \text{kJ mol}^{-1}$
Electron gain	$\text{Cl(g)} + \text{e} \longrightarrow \text{Cl}^-\text{(g)}$	$A = -347\ \text{kJ mol}^{-1}$
	$\text{Na}^+\text{(s)} + \text{Cl}^-\text{(g)} \longrightarrow \text{NaCl(s)}$	$\Delta H = -779\ \text{kJ mol}^{-1}$

This corresponds to a lattice energy ΔU (298 K) of $-774\ \text{kJ mol}^{-1}$.
See also Da.

Le Chatelier's principle

Le Chatelier's principle states that 'if a change occurs in one of the factors such as temperature, pressure or concentration under which a system is in *equilibrium* (q.v.), the system will tend to adjust itself, so as to annul, as far as possible, the effect of that change'.

This principle is a direct consequence of the *second law of thermodynamics* (q.v.) and is applicable to all systems in equilibrium.

Thus, if the pressure of a gas mixture in equilibrium is increased, i.e. the volume occupied is decreased, the reaction will tend to occur which involves a decrease in the number of molecules, and the position of equilibrium will be shifted in that direction. For example, for the system

$$\text{N}_2\text{O}_4\text{(g)} \rightleftharpoons 2\text{NO}_2\text{(g)}$$

at constant temperature, the application of pressure causes a displacement of the equilibrium from right to left. (N.B. At constant temperature, the *equilibrium constant*, q.v., remains constant.)

In a similar manner, if the temperature is increased, that reaction will tend to occur in which heat is absorbed. For example, for the endothermic reaction

$$N_2(g) + O_2(g) \rightleftharpoons 2NO(g)$$

the higher the temperature, the greater will be the equilibrium constant and the greater the concentration of nitric oxide (see *van't Hoff isochore*).

Line integral
See Exact differential.

M

Massieu function
The Massieu function, J (dimensions: m l^2 t^{-2} deg^{-1}; units: J K^{-1}), a possible thermodynamic function which has been superseded by the *maximum work function* (q.v.), is defined by

$$J = -A/T = -U/T + S$$

whence,

$$dJ = \frac{U\,dT}{T^2} + \frac{P\,dV}{T}$$

Maximum work function
See Helmholtz free energy.

Maxwell–Boltzmann distribution
As a result of molecular collisions in the gaseous phase, there is a wide diversity in the velocities among individual molecules at a given instant. Although any value of the velocity, between zero and an infinite velocity, is theoretically possible, the fraction of the total number of molecules with speed greater than the most probable velocity is not large. Since a gas is uniformly distributed throughout the containing vessel, it is obvious that all directions of motion are equally probable.

Out of a total number of molecules N, the Maxwell–Boltzmann law shows that the probability p of finding molecules within the velocity range c and $(c + \mathrm{d}c)$ is

$$p = \frac{\mathrm{d}N/N}{\mathrm{d}c} = 4\pi(m/2\pi kT)^{3/2}\, c^2 \exp(-mc^2/2kT) \tag{M.1}$$

Since, for any specific velocity c the kinetic energy of a gas molecule is $\varepsilon = \frac{1}{2}mc^2$, this may be substituted in equation (M.1) to give the distribution of kinetic energies in a gas:

$$\frac{\mathrm{d}N/N}{\mathrm{d}\varepsilon} = \frac{2\varepsilon^{1/2}}{\pi^{1/2}(kT)^{3/2}} \exp(-\varepsilon/kT) \tag{M.2}$$

The curve of $(\mathrm{d}N/N)/\mathrm{d}c$ or $(\mathrm{d}N/N)/\mathrm{d}\varepsilon$ against c or E becomes broader and less peaked at higher temperatures as the average velocity or energy increases and the distribution about the peak becomes wider (figure M.1).

When the molecular energy of a gas can be expressed as the sum of two squared terms, then the average value of energy is kT and the fraction of

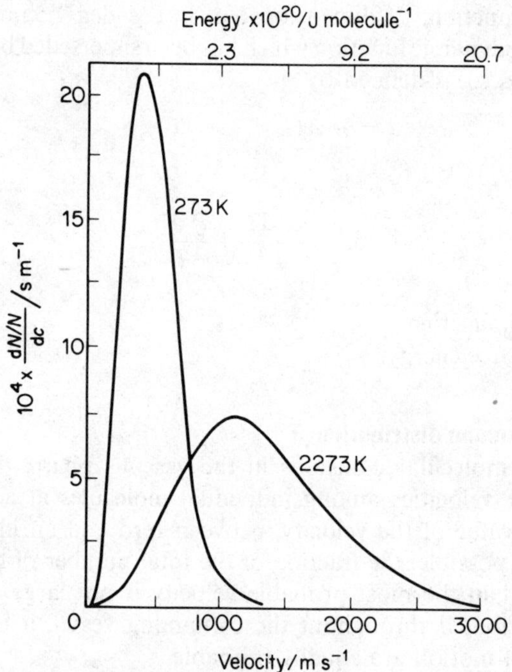

Figure M.1 Distribution of molecular speeds (nitrogen).

molecules possessing a certain energy ε_r or more is given by

$$dN/N \text{ (for } \varepsilon \geq \varepsilon_r) = \exp(-\varepsilon_r/kT) \quad \text{(M.3)}$$

This is the only example where the fractional number of molecules exactly equals the Boltzmann factor without a temperature-dependent term. Equation (M.3) is the most useful form of the distribution law in kinetics and in thermodynamics.

If, however, the energy of the molecule is expressible as the sum of $2s$ squared terms (s potential energy and s kinetic energy terms), the distribution law may be rewritten

$$dN/N \text{ (for } \varepsilon \geq \varepsilon_r) = \frac{(\varepsilon_r/kT)^{s-1}}{(s-1)!} \exp(-\varepsilon_r/kT) \quad \text{(M.4)}$$

and the average energy of the molecule ε is skT. The average energy of 1 mole of molecules of the type under discussion is sRT and the molar heat capacity is sR. A system of this type is sometimes referred to as a system of s feebly coupled oscillators, implying a coupling energy of zero.

From equation (M.4), it is apparent that the number of molecules which have an energy greater than ε_r distributed between all the possible degrees of freedom is much larger than that given by equation (M.3). This can account for the abnormally high rates of reaction often recorded.

See also Ad, P (I).

Maxwell equations

Since the *free energy* (q.v.) function G is a *state function* (q.v.), the exact differential dG is given by

$$dG = V\,dP - S\,dT$$

The application of *Euler's reciprocity relationship* (q.v.) gives

$$\left(\frac{\partial V}{\partial T}\right)_P = \left(\frac{\partial S}{\partial P}\right)_T \quad \text{(M.5)}$$

Applying the same principle to the equation

$$dA = -P\,dV - S\,dT$$

gives

$$\left(\frac{\partial P}{\partial T}\right)_V = \left(\frac{\partial S}{\partial V}\right)_T \quad \text{(M.6)}$$

Similarly, since

Maxwell equations

$$dU = T\,dS - P\,dV$$

then

$$\left(\frac{\partial T}{\partial V}\right)_S = -\left(\frac{\partial P}{\partial S}\right)_V \qquad \text{(M.7)}$$

and

$$dH = T\,dS + V\,dP$$

then

$$\left(\frac{\partial T}{\partial P}\right)_S = -\left(\frac{\partial V}{\partial S}\right)_P \qquad \text{(M.8)}$$

These equations are known as the Maxwell reciprocity relationships. They are useful in deriving other thermodynamic equations, e.g. the *Clausius–Clapeyron equation* (q.v.) from equation (M.6).

See also Dnb, P (I).

Melting point

The normal melting point of a solid is the temperature at which liquid first appears when the solid is heated at atmospheric pressure. During the uniform cooling of a liquid, the temperature remains constant at a particular value, owing to the evolution of the heat of fusion, until all the liquid has been converted into solid. This is the freezing point of the liquid or melting point of the solid; it is an invariant point (see *invariant system*), $f' = 0$, for the *one-component system* (q.v.). The *Clausius–Clapeyron equation* (q.v.) expresses the variation of m.p. with applied pressure.

In a *two-component condensed system* (q.v.) in which compound formation occurs, if the solid can exist in equilibrium with liquid of the same composition, it is said to have a *congruent melting point* (q.v.). On the other hand, if the compound is not very stable and decomposes at temperatures below its m.p., it is said to have an *incongruent melting point* (q.v.). In the latter case, known as meritectic melting, the solid establishes equilibrium with a second solid phase and a liquid of different composition. This can be represented by a chemical reaction in which chemical bonds are broken:

$$\text{Solid I} \rightleftharpoons \text{Liquid} + \text{Solid II}$$

Peritectic melting involves no bond rupture and may be represented:

$$\text{Solid solution } \alpha \rightleftharpoons \text{Solid solution } \beta + \text{Liquid}$$

Since solid solutions are not compounds, this is not a chemical reaction but a phase change.

See also Bo.

Meritectic melting

See Incongruent melting point; Melting point.

Metal extraction

See Ellingham diagram.

Metre

The metre, m, is the basic SI unit of length and is defined as the length equal to 1 650 763.73 wavelengths in vacuum of the radiation corresponding to the transition between the levels $2p_{10}$ and $5d_5$ of the ^{86}Kr atom.

Mixing

See Entropy; Free energy; Ideal solution; Real solution.

Mixtures, constant boiling

See Azeotrope; Binary liquid mixtures.

Molality

The molality, m (units: mol kg^{-1}), of a solution is the concentration expressed as moles of the solute per kilogramme of solvent. A molal solution is prepared by dissolving one-thousandth of the molecular (formula) weight in kg (i.e. formula weight in g) of the solute in 1 kg of solvent.

The mean ionic molality, $m_\pm$ of an electrolyte is defined:

$$m_\pm = (m_+^{\nu+} m_-^{\nu-})^{1/\nu} = m(\nu_+^{\nu+} \times \nu_-^{\nu-})^{1/\nu}$$

See also Concentration units.

Molarity

The molarity of a solution is the concentration expressed as moles of solute per dm^3 (litre) of solution. This is an obsolete unit of concentration, and its use with the SI units is discouraged. A 1 molar solution is equivalent to a concentration of 1 mol dm^{-3}.

See also Concentration units.

Molar volume

Molar volume, V_m (units: m^3 mol^{-1}), is the volume occupied by 1 mole of substance. Numerically, it is equal to the relative molecular mass divided by the density:

$$V_m = V/n = M_r/\rho$$

Thus, for water,

$$V_m = 0.018\,015/(0.99707 \times 10^3) = 18.068 \times 10^{-6}\ \mathrm{m^3\,mol^{-1}}$$

Mole

The mole, n_i, is the amount of substance which contains as many elementary units as there are atoms in 0.012 kg of carbon-12. The elementary unit must be specified, and may be an atom, a molecule, an ion, a radical, an electron, etc. For example, 1 mole of HgCl has a mass equal to 236.04 g; 1 mole of electrons has a mass equal to 5.4860×10^{-4} g; 1 mole of SO_4^{2-} has a mass equal to 96.06 g.

Thus the number of moles in weight w_i g of substance is given by $n_i = w_i/M_r(\mathrm{i})$. The actual number of molecules is obtained by multiplying by the Avogadro constant, i.e. $n_i N_A$.

Mole fraction

Mole fraction (a dimensionless quantity) of the *i*th component, x_i, in a mixture is defined as

$$x_i = \frac{\text{No. of moles of } i\text{th component}}{\text{Total no. of moles of all components}} = \frac{n_i}{\sum_i n_i} = \frac{w_i/M_r(\mathrm{i})}{\sum_i w_i/M_r(\mathrm{i})}$$

The sum of all the mole fractions of all components in a mixture is unity, i.e. $\sum_i x_i = 1$.

See also Concentration units.

Monotropy

Monotropy is exhibited by two allotropic or polymorphic forms when the transition temperature lies above the m.p. of the solid (figure M.2). Only one form is stable at all temperatures up to the m.p.; the other form is metastable. The transformation is irreversible. Examples of monotropic systems are phosphorus (see *phosphorus system*), iodine monochloride and benzophenone.

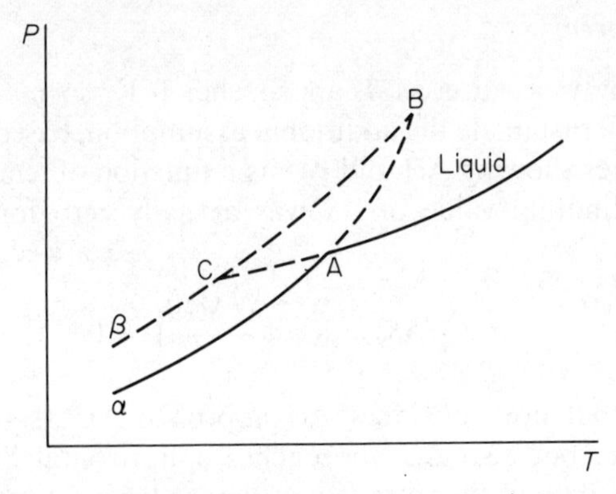

Figure M.2 Vapour pressure–temperature curve for a monotropic system. (B is the transition temperature, A the m.p. of solid α, and C the m.p. of the metastable form solid β.)

N

Natural process
See Spontaneous process.

Nernst distribution law
See Distribution law.

Nernst heat theorem
The trend of ΔG and ΔH towards each other as the temperature decreases, expressed by

$$\lim_{T \to 0} (\Delta G - \Delta H) = 0$$

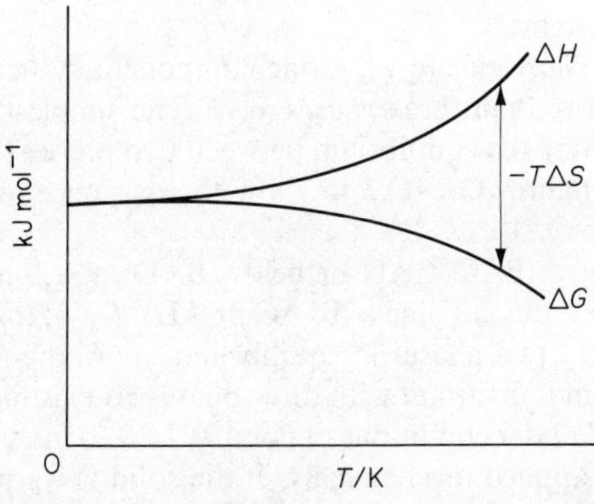

Figure N.1 Variation of the enthalpy and free energy changes with temperature.

may follow simply because, as T approaches 0 K, so must $T\Delta S$ as long as ΔS is finite. Nernst made the additional assumption, based on the appearance of the curves showing ΔH and ΔG as a function of temperature (figure N.1), that the limiting value of ΔS was actually zero for all condensed systems, i.e.

$$\lim_{T\to 0} (-\Delta S) = \lim_{T\to 0} \left(\frac{\partial(\Delta G)}{\partial T}\right)_P = 0$$

This means that not only does ΔG approach ΔH as the temperature approaches zero, but also ΔG approaches a horizontal limiting tangent. In Nernst's postulate of the *third law of thermodynamics* (q.v.) that ΔS and ΔC_p become zero at 0 K for all reactions between condensed phases, there is no reference to the value of the absolute *entropy* (q.v.) at 0 K, although it must be finite or zero.

Numerous experiments have confirmed the postulate, provided it is limited to a single perfectly crystalline substance.

See also Enthalpy; Free energy; and K, Wi.

Newton

The newton, N (dimensions: $\mathrm{m\,l\,t^{-2}}$; units: $\mathrm{N = kg\,m\,s^{-2} = J\,m^{-1}}$), is the SI unit of force, defined as that force which when applied to a body having a mass of 1 kg, gives it an acceleration of 1 metre per second squared. ($\mathrm{N} = 10^5$ dyne.)

O

One-component system

According to the *phase rule* (q.v.), a one-component system in equilibrium can exist in no more than three *phases* (q.v.). The simplest one-component systems are shown by the equilibrium between two phases (figures O.1–O.4) or by combining figures O.1–O.3 to show the existence of three phases in equilibrium (figure O.5).

In any area (above BAC, CAD or below BAD), $f = 2$ and the system is said to be bivariant. On any line (AB, AC or AD), $f = 1$; this is a univariant system consisting of two phases in equilibrium. At A, the *triple point* (q.v.), $f = 0$ and the system is invariant with three phases co-existing in equilibrium.

Consider a solid under conditions of point W ($f = 2$); no vapour is present, so the pressure is applied mechanically. If the solid is warmed isobarically, it will melt at X ($f = 1$) and the pressure is still mechanical. On further

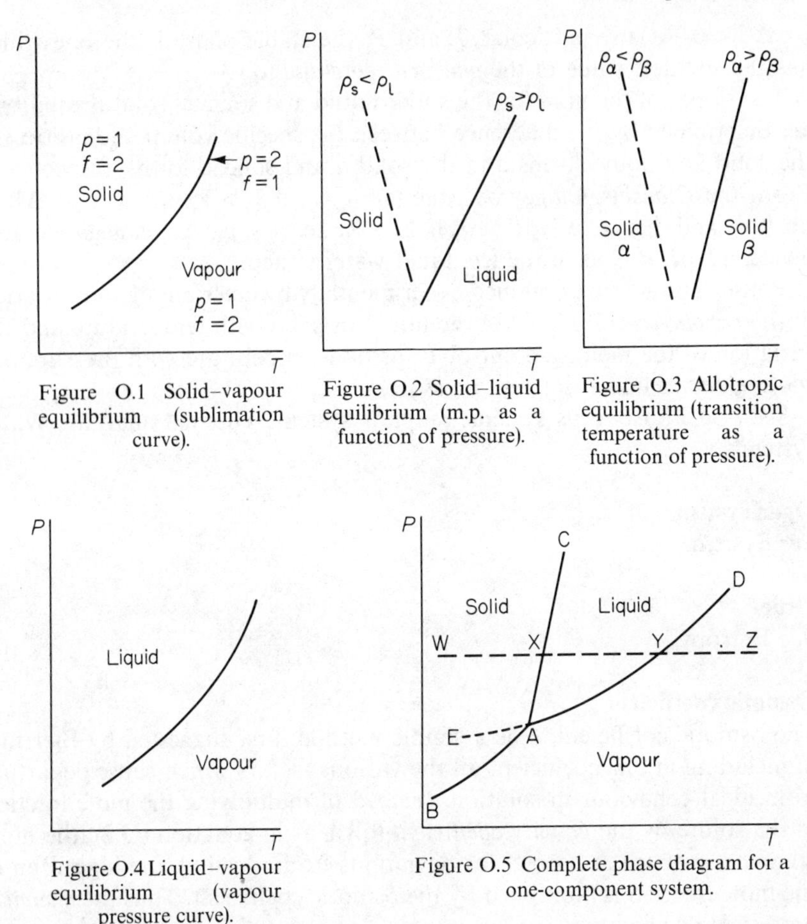

Figure O.1 Solid–vapour equilibrium (sublimation curve).

Figure O.2 Solid–liquid equilibrium (m.p. as a function of pressure).

Figure O.3 Allotropic equilibrium (transition temperature as a function of pressure).

Figure O.4 Liquid–vapour equilibrium (vapour pressure curve).

Figure O.5 Complete phase diagram for a one-component system.

heating, the temperature rises along XY ($f = 2$) until, at Y ($f = 1$), the vapour phase appears (i.e. the liquid boils at the given pressure); the pressure of the system is now produced by the vapour. At point Z ($f = 2$), the system is entirely in the vapour phase. If this operation were conducted at a pressure below that of the triple point A, the liquid phase would not appear; instead there would be direct sublimation from the solid to the vapour phase.

Extension of DA below the triple point to E results in the metastable equilibrium between the supercooled liquid and its vapour. This can be achieved experimentally under carefully controlled conditions. The metastable form at a given temperature always has the higher vapour pressure, since conversion to the form stable at that temperature is spontaneous, i.e. $\Delta G < 0$. At any particular T and P, the stable phase is the one which has the smallest value of the *chemical potential* (q.v.).

The slopes of the lines for the solid–liquid and solid α–solid β equilibria are determined by the difference between the specific volumes (densities) of the solid and liquid forms and the solid α and solid β forms, respectively. From the *Clausius–Clapeyron equation* (q.v.), if $V_l > V_s$, i.e. $\rho_s > \rho_l$, ΔV is positive and, hence, $dT/dP > 0$; if $V_s > V_l$, i.e. $\rho_s < \rho_l$, ΔV is negative and, hence, $dT/dP < 0$, as in the ice-liquid water system.

Phase diagrams are obtained experimentally by determining the variation of the *vapour pressure* (q.v.) of the liquid or solid with temperature and the variation of the melting point or transition temperature with pressure (i.e. cooling curves under different pressures).

See also Phosphorus system; Sulphur system; Thermal analysis; Water system.

Open system
See System.

Order
See Entropy.

Osmotic coefficient
The osmotic coefficient, g, is a simple method, first suggested by Bjerrum, of including in one coefficient all the various factors which cause departure from ideal behaviour in solution. Instead of multiplying the mole fraction of the solute by the *activity coefficient* (q.v.), as in equation (O.2) (the most common practice to correct for deviations from ideality), the logarithm of the mole fraction is multiplied by the osmotic coefficient. Thus the *chemical potential* (q.v.) becomes

$$\mu_i = \mu_i^\ominus + g_i RT \ln x_i \qquad \text{(O.1)}$$

instead of

$$\mu_i = \mu_i^\ominus + RT \ln f_i x_i \qquad \text{(O.2)}$$

For ideal systems, both these equations become

$$\mu_i = \mu_i^\ominus + RT \ln x_i \qquad \text{(O.3)}$$

Comparing equations (O.1) and (O.2), it is apparent that

$$\ln f_i = (g_i - 1) \ln x_i \qquad \text{(O.4)}$$

or

$$g_i \ln x_i = \ln f_i x_i \tag{O.5}$$

If the pure species behaves ideally, then $g_i = 1$ (compare equations O.1 and O.3); hence, g_i approaches a limiting value of 1. Although g_i exists for each species, in any solution only the value of g_A for the solvent A is widely used. Experimental values of g_A (and also f_A) can be found directly from elevation of *boiling point* (q.v.) or depression of *freezing point* (q.v.) of a solvent. In the elevation of the boiling point for an ideal solution, integration of the *Clausius–Clapeyron equation* (q.v.) in the form

$$-\left(\frac{\partial \ln x_A}{\partial T}\right)_P = \frac{L_e}{RT^2}$$

between $x_A = 1$ at the b.p. of the solvent T_0 and x_A at the higher temperature T leads to the usual equation for elevation of b.p.:

$$-\ln x_A = \frac{L_e \, \Delta T_e}{RT_0^2} \tag{O.6}$$

Allowing for deviations from ideal behaviour, these equations become

$$\frac{L_e}{RT^2} = -\left(\frac{\partial(g_A \ln x_A)}{\partial T}\right)_P = -\left(\frac{\partial(\ln f_A x_A)}{\partial T}\right)_P$$

Since g_A, f_A, and x_A are all unity when $T = T_0$, integration between the same limits gives

$$-g_A \ln x_A = -\ln f_A x_A = \frac{L_e \, \Delta T_e}{RT_0^2} \tag{O.7}$$

Thus g_A and f_A can be calculated from observed values of L_e, x_A, T and T_0. Rearranging equation (O.7) gives

$$\Delta T = g_A \frac{RT_0^2}{L_e} \ln(1/x_A) = g_A \, \Delta T^*$$

Thus g is the factor by which the elevation of the b.p. and depression of the f.p. of an ideal solution, ΔT_e^*, must be multiplied to give these quantities for an actual solution of the same concentration.

The reason for naming g the 'osmotic coefficient' stems from the *osmotic pressure* (q.v.) equation:

$$\pi \bar{V}_A = -g_A RT \ln x_A$$

which results when equation (O.1) replaces equation (O.3). By the usual

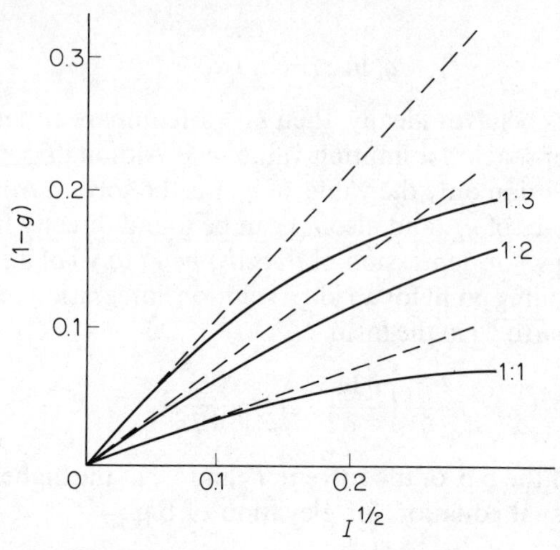

Figure O.6 Osmotic coefficient of water in dilute aqueous electrolyte solutions (dotted lines indicate limiting Debye–Hückel slope).

approximations, when $n_A \gg n_B$, this equation becomes

$$\pi \bar{V}_A = g_A RT$$

The factor $(1 - g)$ measures the departure of a species from ideal behaviour. In dilute solutions of electrolytes (figure O.6),

$$(1 - g) = -\tfrac{1}{3} \ln f_{\pm}$$

and, hence, the limiting Debye–Hückel equation (see *Debye–Hückel activity equation*) in dilute aqueous solution at 298 K can be written

$$(1 - g) = 0.3921\ z_{+} z_{-} I^{1/2}$$

See also I, R & S.

Osmotic pressure

The osmotic pressure, π (dimensions: m l^{-1} t^{-2}; units: N m^{-2}), of a solution is the minimum excess pressure to which it must be subjected to prevent net transfer of solvent through a semipermeable membrane (i.e. a membrane not permeable to solute molecules) separating it from pure solvent. During osmosis, solvent molecules pass from the solvent (or more dilute solution) through the membrane into the more concentrated solution, thereby building up a pressure if the solution is confined.

The osmotic pressure between a dilute solution and solvent is a *colligative property* (q.v.) and depends only on the concentration of the solution. Assuming ideal behaviour, π is related to the v.p. above the solvent A and solution by

$$\pi\,\bar{V}_A = RT\ln(p^{\ominus}/p) = -RT\ln x_A$$

which, assuming the validity of *Raoult's law* (q.v.), becomes

$$\pi\,\bar{V}_A = RT\,x_B = RT\,(n_B/n_A) \qquad (n_A \gg n_B)$$

or

$$\pi = \frac{n_B RT}{n_A\,\bar{V}_A} = \frac{n_B RT}{V} = c_B RT = \frac{w_B RT}{M_r(B)}$$

This equation is only strictly valid at infinite dilution; π for a solution of a non-electrolyte can be represented by a power series in weight per unit volume, w_B;

$$\frac{\pi}{w_B} = \frac{RT}{M_r(B)} + C\,w_B + D\,w_B^2 \ldots$$

Thus the plot of π/w_B against w_B is linear and of intercept $RT/M_r(B)$, from which $M_r(B)$ can be obtained.

The osmotic pressure is related to both the elevation of *boiling point* (q.v.) and the depression of *freezing point* (q.v.) of a solvent on the addition of a solute.

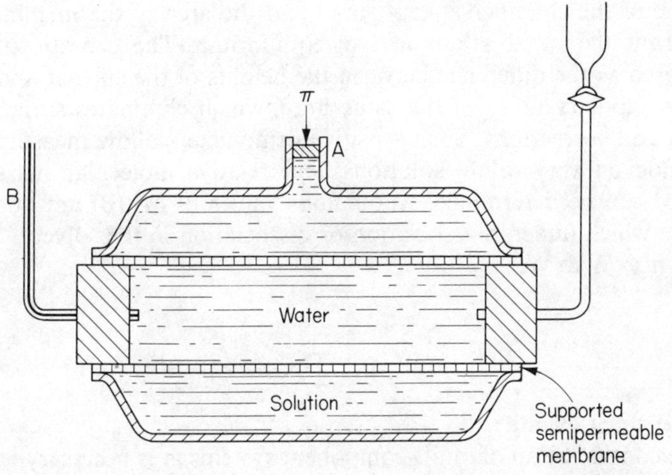

Figure O.7 Osmotic pressure apparatus of Berkeley and Hartley.

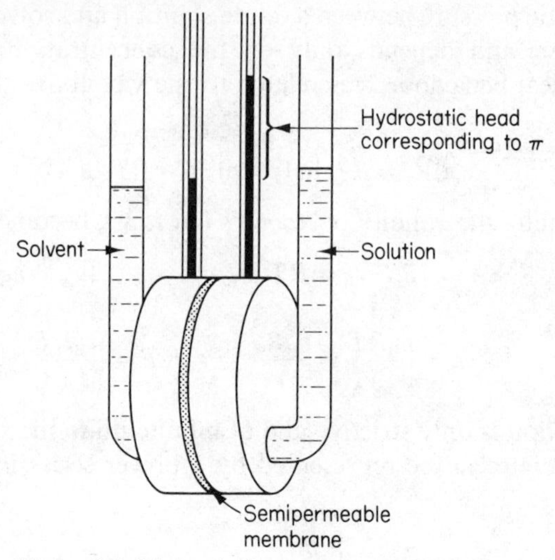

Figure O.8 Schematic diagram of osmometer.

Accurate measurements of large osmotic pressures require strong membranes such as those of collodion, swollen cellophane or the supported copper ferrocyanide used by early workers. In the Berkeley and Hartley method, pressure is applied at A (figure O.7) until the water level in the capillary tube at B remains constant; the applied pressure is equal to the osmotic pressure. In commercially available osmometers (figure O.8), the volume of the chamber is kept small and the area of the membrane large to permit the rapid attainment of equilibrium. The osmotic pressure is measured as the difference between the heights of the solvent and solution in two capillary tubes of the same bore, which eliminates surface tension errors and corrections. Such sensitive osmometers allow measurements to be made on very dilute solutions, and relative molecular masses up to 5×10^5 can be determined. Anomalous values of M_r (B) are obtained for solutes which undergo association or dissociation in the solvent.

See also W & W.

P

Partial molar quantity

In the consideration of multi-component systems, it is necessary to consider the change of a thermodynamic property due to changes in the composition as well as to changes in the normal variables, e.g. P, V, T. A partial molar quantity is the increase in any extensive property of a system when 1 mole of one component is added to an amount of the system so large that the composition and other intensive properties do not change appreciably when the addition is made. If the composition and amount of materials are variables, the extensive property, J, of the whole system is a function of n_i, the number of moles of each component in addition to such variables as T and P, i.e.

$$J = f(T, P, n_A, n_B, \dots n_i)$$

Differentiating at constant T and P gives

$$\mathrm{d}J = \left(\frac{\partial J}{\partial n_A}\right)_{T,P,n_j} \mathrm{d}n_A + \left(\frac{\partial J}{\partial n_B}\right)_{T,P,n_j} \mathrm{d}n_B + \dots$$

$$= \bar{J}_A\, \mathrm{d}n_A + \bar{J}_B\, \mathrm{d}n_B \dots = \sum_i \bar{J}_i\, \mathrm{d}n_i$$

where $\bar{J}_i = (\partial J/\partial n_i)_{T,P,n_j}$ is known as the partial molar property of the ith component (n_j indicates that the numbers of moles of all components except the one indicated inside the partial derivative coefficient remain constant).

Integration at constant T, P and constant composition gives

$$J = \bar{J}_A n_A + \bar{J}_B n_B + \dots = \sum_i \bar{J}_i n_i$$

A change of composition (except for ideal binary mixtures) results in changes in $\bar{J}_i$ as well as n_i; hence, at constant T and P,

$$\mathrm{d}J = \bar{J}_A \mathrm{d}n_A + n_A \mathrm{d}\bar{J}_A + \bar{J}_B \mathrm{d}n_B + n_B \mathrm{d}\bar{J}_B \dots$$

Hence,

$$n_A \mathrm{d}\bar{J}_A + n_B \mathrm{d}\bar{J}_B \dots = \sum_i n_i \mathrm{d}\bar{J}_i = 0$$

For two-component systems,

$$\mathrm{d}\bar{J}_A = -\frac{n_B}{n_A}\,\mathrm{d}\bar{J}_B$$

This equation, which is one form of the *Gibbs–Duhem equation* (q.v.), may be integrated from infinitely dilute solution ($n_B = 0$) to give

$$\int_{(n_B/n_A)=0}^{(n_B\ n_A)} \mathrm{d}\bar{J}_A = \bar{J}_A - \bar{J}_A^{\ominus} = -\int_{(n_B/n_A)=0}^{(n_B/n_A)} \frac{n_B}{n_A}\,\mathrm{d}\bar{J}_B$$

where $\bar{J}_A^{\ominus}$ is the partial molar quantity of the solvent at infinite dilution, i.e. in the pure state; thus, for water at 298 K, $\bar{V}^{\ominus} = 18.0 \times 10^{-6}$ m^3 mol^{-1}.

Partial molar quantities, which are intensity factors, can be defined for any extensive *state function* (q.v.); the most important partial molar quantities are *partial molar volume* (q.v.), $\bar{V}_i$, partial molar *enthalpy* (q.v.), $\bar{H}_i$, and partial molar free energy, $\bar{G}_i$ (commonly known as the *chemical potential* (q.v.), μ_i).

All thermodynamic relations can be applied to partial molar quantities. For example,

$$\left(\frac{\partial \bar{G}_A}{\partial P}\right)_T = \left(\frac{\partial \mu_A}{\partial P}\right)_T = \bar{V}_A$$

$$\left(\frac{\partial \mu_A}{\partial T}\right)_P = -\bar{S}_A$$

Partial molar quantities may be calculated if values of J are known for the whole system. Several methods are available; the simplest methods are described under *partial molar volume*.

See also K, L & R.

Partial molar volume

Partial molar volume of the ith component of a mixture is defined by

$$\bar{V}_I = \left(\frac{\partial V}{\partial n_i}\right)_{T,P,n_j}$$

Such a function is necessary, since the volume of a solution is not, in general, the sum of the volumes of the individual components, owing to the change in character of the molecules on mixing. For an ideal solution, the volume of a component in solution is identical with its volume in the pure state. If ethanol and water formed an ideal solution, the molar volume of a solution in which the mole fraction of each component is 0.5 would be given by

$$V_m = 0.5(18.0 \times 10^{-6}) + 0.5(58.6 \times 10^{-6}) = 38.3 \times 10^{-6} \text{ m}^3$$

where the molar volumes of water and ethanol are 18.0×10^{-6} and 58.6×10^{-6} m^3 mol^{-1}, respectively; when such a solution is prepared, its actual volume is 36.9×10^{-6} m^3.

In figure P.1 $(\partial V_B/\partial n_B)_{T,P,n_A}$ is constant, i.e. no matter what the concentration of solute (over the range plotted), the addition of further amounts increases the volume by equal amounts per mole; under these conditions, $\bar{V}_B$ is the effective volume per mole. In practice, this effective volume of

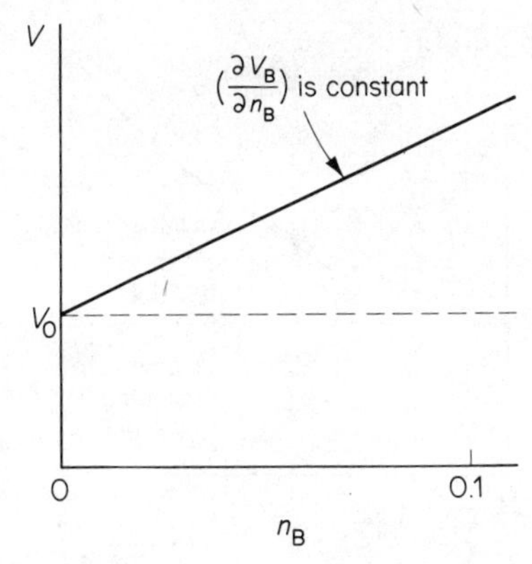

Figure P.1 Linear dependence of volume on concentration (ideal mixture).

Table P.1

Solute	Solvent	$10^6\ \bar{V}_B/m^3\ mol^{-1}$ In solution	In pure state
Glycolamide	Water	56.2	54.0 (s)
Iodine	Perfluoro-n-heptane	100	51.0 (s) 59.0 (supercooled liquid)
Hydrogen (101 325 N m^{-2})	Water	26	25 000 (g)
	Ether	50	
	Acetone	38	
NaCl	Water	16.4	27 (s)
Na_2CO_3	Water	−6.7	42 (s)

solute in solution depends on the solvent and may be very different from that in the pure state, as shown in table P.1.

Dependence of the total volume on n_B usually follows the curve (figure P.2) with continuously changing slope; i.e. the volume does not change by equal increments when small quantities of solute are added. Hence, in quoting $\bar{V}_B$, the concentration n_B must also be specified.

The total volume of a solution is given by

$$V = \bar{V}_A n_A + \bar{V}_B n_B$$

whence,

$$n_A\, d\, \bar{V}_A + n_B\, d\, \bar{V}_B = 0$$

This is a form of the *Gibbs–Duhem equation* (q.v.).

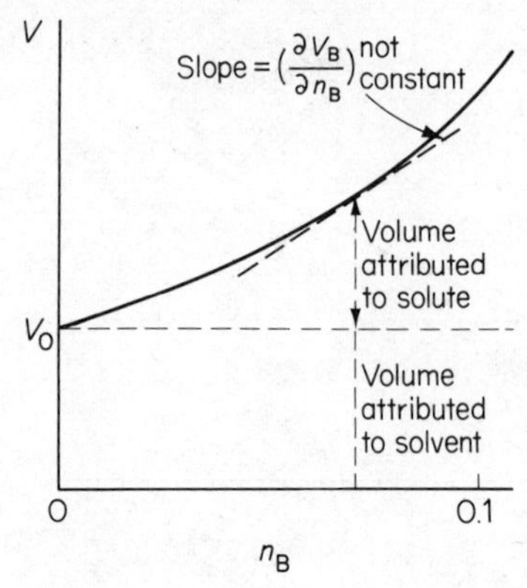

Figure P.2 Non-linear dependence of volume on concentration (non-ideal system).

Determination of partial molar volumes of solute B

Partial molar volumes of solute B can be determined in the following ways.

(1) Plot the volume of solution for varying concentrations of solute against m_B and determine the slope at different values of m_B. This is a poor method, because the actual volumes are difficult to measure accurately.

(2) From a knowledge of the density of solutions of different concentrations. The apparent molar volume of solute ϕ_B is defined

$$\phi_B = \frac{V - n_A V_0}{n_B}$$

where V_0 is the molar volume of pure solvent; hence,

$$V = n_B \phi_B + n_A V_0$$

$$\bar{V}_B = \left(\frac{\partial V}{\partial n_B}\right)_{T,P,n_A} = \phi_B + n_B \left(\frac{\partial \phi_B}{\partial n_B}\right)_{T,P,n_A}$$

The partial molar volume of the solvent is given by

$$\bar{V}_A = \frac{V - n_B \bar{V}_B}{n_A} = \frac{1}{n_A}\left\{n_A V_0 - n_B^2 \left(\frac{\partial \phi_B}{\partial n_B}\right)_{T,P,n_A}\right\}$$

The apparent molar volume of the solute can be calculated from the density:

$$\phi_B = \frac{1}{n_B}\left\{\frac{n_A M_r(\mathrm{A}) + n_B M_r(\mathrm{B})}{\rho} - n_A V_0\right\}$$

If n_A is the number of moles of solvent and n_B the number of moles of solute in 1 kg of solvent, then $n_B = m$ (the molality of the solution) and

$$\phi_B = \frac{1}{m}\left\{\frac{1 + mM_r(B)}{\rho} - \frac{1}{\rho_0}\right\}$$

where ρ_0 is the density of the pure solvent. Thus, from density determinations, ϕ_B can be calculated at different values of the molality, and, from the ϕ_B–molality curve, the slope $(\partial\phi_B/\partial n_B)$ at different molalities and, hence, $\bar{V}_A$ and $\bar{V}_B$ can be calculated.

(3) Analytically. The total volume can often be expressed as an algebraic function of the concentration: e.g., for sodium chloride at 298 K and atmospheric pressure, the volume $(10^6 V/m^3)$ can be expressed in terms of the molality:

$$V = 1001.38 + 16.6253\, m_B + 1.7738\, m_B^{3/2} + 0.1194\, m_B^2$$

whence,

$$\bar{V}_B = \left(\frac{\partial V}{\partial n_B}\right)_{T,P,n_A} = 16.6253 + 2.6607\, m_B^{1/2} + 0.2388\, m_B$$

The partial molar volume of the solvent is evaluated by using the equation (see *partial molar quantity*)

$$\bar{V}_A - V_A^\ominus = -\int_0^{m_B} \frac{m_B}{m_A}\, d\,\bar{V}_B = -\int_0^{m_B} \frac{m_B}{m_A}\,(1.3304\, m_B^{-1/2} + 0.2388)\, dm_B$$

Since $V_A^\ominus = \dfrac{M_r(A)}{\rho_0}$

$$\bar{V}_A = 18.08 - 0.015\,997\, m_B^{3/2} - 0.002\,15\, m_B^2$$

See also K.

Partially miscible liquids

When two dissimilar liquids, which differ in polarity or extent of association, are mixed, the forces of attraction between A and B (A–B) are usually less than those between A–A and B–B; under these conditions the liquids tend to separate, the escaping tendency is increased and, hence, a positive deviation from *Raoult's law* (q.v.) is observed. If the forces of attraction between like molecules are very much greater than those between unlike molecules, then the liquid phase separates into two layers; these are saturated solutions of A in B and of B in A in a partially miscible system. Since the *vapour pressure* (q.v.) is a function of the temperature, the range of miscibility depends on the temperature. Such systems are usually treated at constant (normally

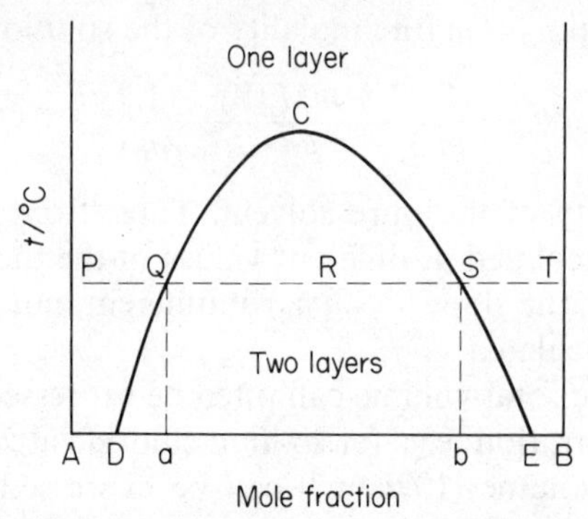

Figure P.3 *T*–composition curve for partially miscible liquid system.

atmospheric) pressure, and the phase diagrams are constructed in terms of the remaining variables of temperature and composition. For a system separated into two layers, the application of the *phase rule* (q.v.) gives

$$f = c - p + 2 = 2 - 2 + 2 = 2$$

but, since the pressure is fixed, the temperature alone will completely define the system; at a given temperature, the compositions (a and b) of the two liquid layers Q and S in equilibrium are fixed and independent of the relative amounts of the two phases (figure P.3). These liquid layers are known as conjugate solutions. Increase in temperature usually results in an increase in the mutual solubility of the two components until, at temperatures greater than the critical solution temperature or consolute temperature, C, the system is homogeneous at all compositions. The point C is the superimposition of the ends of the solubility curves DC and EC, and, for a given system, represents a fixed temperature and composition. Addition of B to A can be considered at constant temperature along the line PQRST:

(1) over the range PQ, a true solution of B in A exists;

(2) beyond Q towards S, two layers separate out, of compositions a and b, such that, at point R,

$$\frac{\text{amount of layer of composition a}}{\text{amount of layer of composition b}} = \frac{\text{RS}}{\text{QR}};$$

(3) over the range ST, a true solution of A in B exists.

Some systems show increasing mutual solubility with increasing tempe-

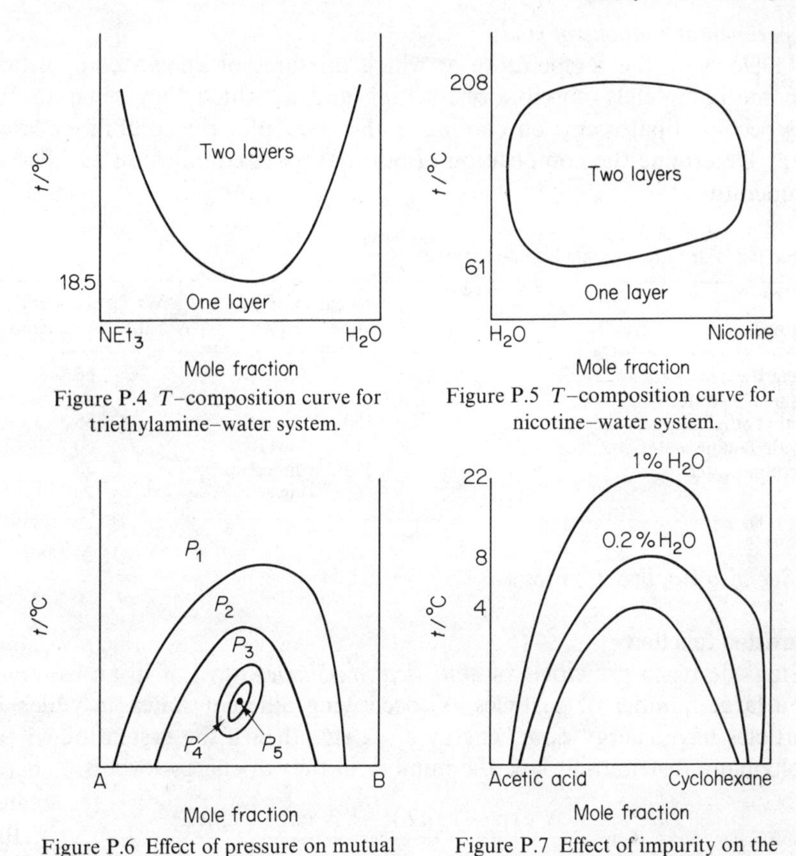

Figure P.4 T–composition curve for triethylamine–water system.

Figure P.5 T–composition curve for nicotine–water system.

Figure P.6 Effect of pressure on mutual solubility curve ($P_5 > P_4 > P_3 > P_2 > P_1$).

Figure P.7 Effect of impurity on the critical solution temperature.

rature, e.g. phenol/water; others show decreasing mutual solubility with increasing temperature, e.g. triethylamine/water (figure P.4); while others show a closed solubility curve with an upper and a lower consolute temperature, e.g. nicotine/water (figure P.5).

An increase in pressure lowers (raises) the upper (lower) consolute temperature (figure P.6). The consolute temperature is very sensitive to the presence of a third substance (figure P.7) and may be used as a test of the purity of a liquid (e.g. the effect of adding water to the acetic acid/cyclohexane system). In general, the consolute temperature is raised if the impurity is soluble in only one liquid and lowered if soluble in both; the solubility curve becomes progressively distorted.

Partially miscible liquids

Experimental methods of study
(1) Determine the temperature at which mixtures of known composition become completely miscible on heating and at which they separate out (i.e. become opalescent) on cooling, and, hence, plot the solubility curves.
(2) Determine the composition of both layers in equilibrium at different temperatures.

Table P.2. Partially miscible liquid mixtures

Liquids	Critical solution temperature/°C	Wt. % of water at critical soln. temp.
Phenol/water	65.9 (upper)	66
Butan-2-ol/water	113.5 (upper)	70
Methyl ethyl ketone/water	150 (upper)	55
Triethylamine/water	18 (lower)	60
Nicotine/water	208 (upper)	70
	61 (lower)	70

See also Bo, F & C, F & J.

Partition function
Using the basic principles of *statistical mechanics* (q.v.), if in an assembly of a large number of particles, N, occupying different states, in which n_0 particles have energy ε_0, n_1 energy $\varepsilon_1 \ldots$, etc., then if the system obeys the Boltzmann distribution law, the number in the rth energy level is

$$n_r = \frac{N \exp(-\varepsilon_r/kT)}{\sum_{r=0}^{r=\infty} \exp(-\varepsilon_r/kT)} = \frac{N \exp(-\varepsilon_r/kT)}{Q} \qquad \text{(P.1)}$$

where Q, the partition function or state sum (Zustandsumme), is defined as

$$Q = \sum_{r=0}^{r=\infty} \exp(-\varepsilon_r/kT) \qquad \text{(P.2)}$$

The partition function thus determines how the particles are partitioned among the different energy levels. Frequently there are several energy states with almost equal energies; if there are g_r states of energy very close to ε_r, then it is usual to regard them as a single degenerate state of statistical weight g_r; then

$$Q = \sum_{r=0}^{r=\infty} g_r \exp(-\varepsilon_r/kT) \tag{P.3}$$

and the number of particles with energy ε_r becomes

$$n_r = \frac{N\, g_r \exp(-\varepsilon_r/kT)}{Q} \tag{P.4}$$

The mathematical probability of the system is defined as the number of distinguishable ways in which the particles can be assigned to the different energy levels, and is determined solely by the largest term ($t_{\max}$) when N is very large, i.e.

$$W = t_{\max} = \frac{N!}{\prod_r n_r!} \prod_r (g_r)^{n_r}$$

Hence,

$$\begin{aligned} \ln t_{\max} &= \ln N! - \sum_r \ln n_r! + \sum_r n_r \ln g_r \\ &= N \ln N - \sum_r n_r \ln n_r + \sum_r n_r \ln g_r \end{aligned} \tag{P.5}$$

(using the Stirling approximation that, when N is large, $\ln N! = N \ln N - N = N \ln N/\mathrm{e}$).

Relation of partition function to thermodynamic functions

$$U = \sum_r n_r \varepsilon_r = N_A k T^2 \left(\frac{\partial \ln Q}{\partial T}\right)_V \tag{P.6}$$

$$\begin{aligned} S = k \ln W = k \ln t_{\max} &= N_A k \ln Q + N_A k T \left(\frac{\partial \ln Q}{\partial T}\right)_V \\ &= N_A \ln Q + U/T \end{aligned} \tag{P.7}$$

$$A = -N_A k T \ln Q \tag{P.8}$$

$$P = -\left(\frac{\partial A}{\partial V}\right)_T = N_A k T \left(\frac{\partial \ln Q}{\partial V}\right)_T \tag{P.9}$$

$$G = A + PV = -N_A k T \ln Q + N_A k T V \left(\frac{\partial \ln Q}{\partial V}\right)_T \tag{P.10}$$

If the assembly is a perfect gas where the molecules have no identifiable positions, then $t_{\max}$ is smaller by a factor of $N_A!$; hence, the functions become

Partition function

$$U = \tfrac{3}{2} N_A kT \tag{P.6a}$$

$$S = N_A k \ln Q + U/T - k \ln N_A! = N_A k \ln Q - N_A k \ln N_A + N_A k + U/T$$
$$= N_A k \ln Qe/N_A + U/T \tag{P.7a}$$

$$A = -N_A kT \ln Q + kT \ln N_A! = -N_A kT (\ln Q/N_A + 1) \tag{P.8a}$$

$$G = -N_A kT \ln Q/N_A \tag{P.10a}$$

$$C_V = \left(\frac{\partial U}{\partial T}\right)_V = 2N_A kT \left(\frac{\partial \ln Q}{\partial T}\right)_V + N_A kT^2 \left(\frac{\partial^2 \ln Q}{\partial T^2}\right)_V \tag{P.11}$$

$$\text{Chemical potential per molecule} = \frac{\mu}{N_A} = -kT \ln Q + kT \ln N_A$$
$$= -kT \ln Q/N_A$$

Calculation of the value of the partition function

$$Q = g_0 \exp(-\varepsilon_0/kT) + g_1 \exp(-\varepsilon_1/kT) + \ldots g_r \exp(-\varepsilon_r/kT)$$

If the terms are arranged in increasing order of ε, they progressively decrease, and, for a given temperature, there is a point beyond which the remaining terms make a negligible contribution to Q and can be ignored. The numerical value of Q depends on the spacing of the energy levels. For close spacings, there will be many terms before the terms become negligible and Q is large; however, if the spacings are wide, only a few terms need be counted and Q is small. For gases, the energy states are close and Q is large; while for solids, the energy states are widely spaced and Q is small. Q depends on temperature, each term increasing with increase of temperature.

The energy of a molecule is usually the sum of the contributions from energies of different kinds, each of which may be at any one of a number of different quantised levels, viz. translational, ε_t, rotational, ε_r, vibrational, ε_v, and electronic, ε_e. If the levels of these separate contributions are not affected by the other types of energy present, then

$$U = \sum \varepsilon_i = \varepsilon_t + \varepsilon_r + \varepsilon_v + \varepsilon_e$$

and, hence,

$$Q_{total} = \sum \exp(-U/kT)$$
$$= \sum \exp(-\varepsilon_t/kT) \times \sum \exp(-\varepsilon_r/kT) \times \sum \exp(-\varepsilon_v/kT) \times \sum \exp(-\varepsilon_e/kT)$$

$$= Q_t \times Q_r \times Q_v \times Q_e \tag{P.12}$$

where Q_t, Q_r, Q_v and Q_e are the partition functions for translation, rotation, vibration and electronic energy levels, respectively. In this manner, Q_{total} can be calculated for quite complex molecules. This method is known as factorisation.

Since Q is factorisable, it is apparent that the terms in $\ln Q$ are separable. Thus

$$U = N_A kT^2 \left(\frac{\partial \ln Q_{total}}{\partial T}\right)_V = N_A kT^2 \left(\frac{\partial}{\partial T}(\ln Q_t + \ln Q_r + \ln Q_v + \ln Q_e)\right)_V$$

$$= N_A kT^2 \left(\frac{\partial \ln Q_t}{\partial T}\right)_V + N_A kT^2 \left(\frac{\partial \ln Q_r}{\partial T}\right)_V + N_A kT^2 \left(\frac{\partial \ln Q_v}{\partial T}\right)_V + N_A kT^2 \left(\frac{\partial \ln Q_e}{\partial T}\right)_V \tag{P.13}$$

Hence, the translational, rotational, vibrational and electronic contributions to the total energy, U, can be evaluated. Similarly with other functions, e.g. S. This provides insight into the magnitudes of such functions.

Translational partition function This can also be factorised into three components of motion along three axes at right angles to one another: $Q_t = Q_x Q_y Q_z$. According to the principles of quantum mechanics, the number of permissible energy levels in gases, although large, is finite and depends on the dimensions of the containing vessel. For one degree of freedom along the x-axis, allowable solutions of the Schrödinger equation are

$$\varepsilon_r = \tfrac{1}{2} m u_x^2 = r^2 h^2 / 8mx^2 \tag{P.14}$$

where r is an integer, u_x the component velocity along the x-axis and x the distance between the walls. Substituting ε_r from equation (P.14) into equation (P.2) gives

$$Q_x = \sum_{r=0}^{r=\infty} \exp(-r^2 h^2 / 8mkTx^2) = \int_0^\infty \exp(-r^2 h^2/8mkTx^2)\,\mathrm{d}r$$

$$= \frac{(2\pi mkT)^{1/2} x}{h} \qquad \left(\text{N.B.} \int_0^\infty \exp(-r^2)\,\mathrm{d}r = \tfrac{1}{2}\pi^{1/2}\right)$$

Thus,

Partition function

$$Q_t = \frac{(2\pi mkT)^{3/2}\, xyz}{h^3} = \frac{(2\pi mkT)^{3/2}\, V}{h^3}$$

where V is the total volume of the container.

For a gas of molecular mass M_r, $Q_t = 1.880 \times 10^{20}(M_r T)^{3/2}\, V$; for hydrogen at 298.15 K, $Q_t = 2.77 \times 10^{24}\, V$.

Rotational partition function (1) the allowed rotational energy levels for diatomic molecules or any linear polyatomic molecules are given by

$$\varepsilon_r = J(J+1)\, h^2/8\pi^2 I$$

where I is the moment of inertia and J the rotational quantum number $J = 0, 1, 2, 3\ldots$). The rotational energy level corresponding to a given value of J has a multiplicity of $2J + 1$; i.e. the number of ways of distributing J quanta between two axes of rotation is $2J + 1$ for every case except $J = 0$, when there are two alternatives.

$$Q_r = \sum_{J=0}^{J=\infty} g_J \exp(-\varepsilon_J/kT) = \sum_{J=0}^{J=\infty} (2J+1)\exp(-J(J+1)h^2/8\pi^2 IkT)$$

$$= 8\pi^2 IkT/h^2$$

In homonuclear molecules (e.g. $^{14}N^{14}N$), only all the odd or all the even J values are allowed, depending on the symmetry properties. If the nuclei are different (e.g. $^{14}N^{15}N$, HCl), there are no restrictions on the J values and the *symmetry number* (q.v.), σ, is introduced:

$$Q_r = 8\pi^2 IkT/h^2\sigma = T/\theta_r\sigma$$

where $\theta_r (= h^2/8\pi^2 Ik)$ is known as the characteristic temperature of rotational motion. At any temperature below θ_r, the rotational motion is frozen out and makes no contribution to thermodynamic quantities.

(2) For non-linear polyatomic molecules, it can be shown in a similar manner that

$$Q_r = 8\pi^2(8\pi^3 I_A I_B I_C)^{1/2}(kT)^{3/2}/h^3\sigma$$

where I_A, I_B and I_C are the three principal moments of inertia of the molecule.

Vibrational partition function The vibrational energy states for a simple linear harmonic vibrator are

$$\varepsilon_v = (v + \tfrac{1}{2})h\nu$$

where ν is the frequency of vibration, $\frac{1}{2}h\nu$ the zero point energy and v the vibrational quantum number ($v = 1, 2, 3 \ldots$). The quantum state corresponding to a quantum number v is at an energy $vh\nu$ above the state when $v = 0$; hence, the partition function for each vibrational degree of freedom is

$$Q_v = \sum_{v=0}^{v=\infty} \exp(-vh\nu/kT) = \frac{1}{1-\exp(-h\nu/kT)} = \frac{1}{1-\exp(-\theta_v/T)}$$

where θ_v $(= h\nu/k)$, known as the characteristic temperature of vibration, is generally large compared with room temperature, so most gas molecules are vibrationally unexcited. The quantum of vibrational energy $h\nu \gg kT$.

Electronic partition function The Schrödinger equation may be solved in principle for electronic states. For most molecules, the excited electronic levels lie so far above the ground state compared with kT that all the molecules can be assumed to be in the ground state. Thus Q_e is simply the degeneracy of the electronic ground state, g_0:

$$Q_e = \sum_{\substack{\text{electronic}\\ \text{states}}} \exp(-\varepsilon_e/kT) = g_0$$

The electronic contribution, in a few diatomic molecules, is due to the presence of two separate electronic energy levels lying very close together; e.g., for NO, the ground state is a multiple for which the lowest energy gap is about 0.015 eV, while, for O_2, there are two distinct energy levels separated by 0.97 eV.

The spacing of energy levels near the ground state varies from about 4×10^{-21} J mol^{-1} for translation to 40–400 J mol^{-1} for rotation, 4–40 kJ mol^{-1} for vibration and 40–400 kJ mol^{-1} for electronic energy. Even at low temperatures, translational energy differences are so small compared with RT that translational degrees of freedom are excited to their classical values. Thus their contributions to most thermodynamic functions may be calculated from the simple kinetic theory. The energies of permissible energy levels (rotation, vibration and electronic) are revealed from spectroscopic measurements, and, hence, their contributions to the thermodynamic functions can be calculated from spectroscopic data and added to give the total function (see table P.3).

At 0 K, the translational energy vanishes and the energy of the system is the unchanging value U_0 due to the masses of the particles, nuclear

Table P.3. Thermodynamic functions for ideal gases

Function		Translational*	Internal (i.e. rotational, vibrational, electronic)
$A_T - A_0$	$= A - U_0$	$= -RT[1 + \ln KV(M_r T)^{3/2}]$	$- RT \ln Q_i$
$G_T - G_0$	$= (A - U_0) + PV$	$= -RT \ln KV(M_r T)^{3/2}$	$-RT \ln Q_i$
S_T	$= -(\partial A/\partial T)_V$	$= R[\frac{5}{2} + \ln KV(M_r T)^{3/2}]$	$+R\left[\ln Q_i + T\left(\frac{\partial \ln Q_i}{\partial T}\right)_V\right]$
$U_T^\ominus - U_0^\ominus$	$= (A - U_0) + TS$	$= \frac{3}{2}RT$	$+RT^2\left(\frac{\partial \ln Q_i}{\partial T}\right)_V$
$H_T^\ominus - H_0^\ominus$	$= (U - U_0) + PV$	$= \frac{5}{2}RT$	$+RT^2\left(\frac{\partial \ln Q_i}{\partial T}\right)_V$
$C_V^\ominus$	$= (\partial U/\partial T)_V$	$= \frac{3}{2}R$	$+RT\left[2\left(\frac{\partial \ln Q_i}{\partial T}\right)_V + T\left(\frac{\partial^2 \ln Q_i}{\partial T^2}\right)_V\right]$
$C_p^\ominus$	$= (\partial H/\partial T)_P$	$= \frac{5}{2}R$	$+RT\left[2\left(\frac{\partial \ln Q_i}{\partial T}\right)_V + T\left(\frac{\partial^2 \ln Q_i}{\partial T^2}\right)_V\right]$

$* K = \frac{(2\pi k)^{3/2}}{h^3 N_A^{5/2}}$.

energy and zero-point electronic energy. Since PV and TS are zero for an ideal gas, $A_0 = G_0 = H_0 = U_0$.

From $S_T^\ominus$ and the values of $H_T^\ominus - H_0^\ominus$, the frequently tabulated *free energy function* (q.v.) $(G_T^\ominus - H_0^\ominus)/T$ may be calculated:

$$\frac{G_T^\ominus - H_0^\ominus}{T} = \frac{H_T^\ominus - H_0^\ominus}{T} - S_T^\ominus$$

The contribution for translational energy/J K^{-1} mol^{-1} is

$$(G_T^\ominus - H_0^\ominus)/T = 30.472 - 19.144(\tfrac{3}{2}\log M_r + \tfrac{5}{2}\log T)$$

while the contribution from the internal degrees of freedom is $-R \ln Q_i$.

See also G & S, Kn, N, R, W, Wi.

Partition law
See Distribution law.

Peritectic melting
See Melting point.

pH

The *pH*† of a solution is formally defined as the negative logarithm (to base 10) of the hydrogen ion *activity* (q.v.) of the solution:

$$pH = -\log a(H_3O^+)$$

Except in fairly concentrated acid solutions, hydrogen ion activities are nearly equal to the hydrogen ion concentrations; hence, for normal working conditions,

$$pH = -\log c(H_3O^+)$$

in which the concentration is expressed in mol dm^{-3}. Since $a(H_3O^+)a(OH^-) = 10^{-14}$ mol^2 dm^{-6} in any dilute solution, at neutrality $a(H_3O^+) = a(OH^-) = 10^{-7}$ mol dm^{-3}; hence, the pH value of a neutral solution = 7.0. Acid solutions have pH < 7 and alkaline solutions have pH > 7. For a 0.01 mol dm^{-3} solution of a strong acid, pH = 2; for a weak acid, the pH depends on the concentration and on the *dissociation constant* (q.v.); thus $pH = \frac{1}{2}pK_a - \frac{1}{2}\log c$.

See also D & J.

Phase

A phase is a homogeneous, physically distinct and mechanically separable part of a system. Each phase is separated from other phases by a physical boundary. Thus ice/water/water vapour are three phases; any number of gases mixing in all proportions consitute one phase; a saturated solution is a three-phase system of solution, undissolved solid and vapour; the equilibrium

$$CaCO_3(s) \rightleftharpoons CaO(s) + CO_2(g)$$

is a three-phase system of two solids and one gas.

A system consisting of one phase is homogeneous and a system of more than one phase is heterogeneous.

At any particular temperature and pressure, the stable phase is the one in which the *component* (q.v.) has the smallest value of the *chemical potential* (q.v.).

Phase equilibria

See One-component system; Phase rule; Two-component system; Three-component system.

Phase rule

The phase rule, formulated by Gibbs, provides a general relationship between the number of *degrees of freedom* (q.v.), f, the number of *components* (q.v.), c, and the number of *phases* (q.v.), p of a system in a state of equilibrium:

$$p + f = c + 2$$

The law applies to all macroscopic systems which are in a state of heterogeneous equilibrium (either chemical or physical) and which are influenced only by changes of temperature, pressure and concentration. The phase rule takes no account of molecular complexity. It is generally used to specify the number of independent variables involved when a system is in equilibrium.

$$\begin{aligned} f &= \text{Total number of variables} - \text{Number of known variables} \\ &= p(c-1) + 2 \qquad\qquad\qquad - c(p-1) \\ &= c - p + 2 \end{aligned}$$

A total of $(c - 1)$ concentration terms are required to completely define a system of c components in one phase; thus $p(c - 1)$ terms are required to completely define p phases; and the number 2 accounts for the variables temperature and pressure. Since the *chemical potential* (q.v.) of a component is the same in all phases in equilibrium, there are thus $(p - 1)$ independent equations for one component, and, for c components, there are thus $c(p - 1)$ known variables. Examples are listed in table P.4.

Table P.4

System	c	p	f
Gas in cylinder	1	1	2 (T and P)
Liquid in equilibrium with vapour	1	2	1 (T or P)
Solid/liquid/vapour	1	3	0
$CaCO_3(s) \rightleftharpoons CaO(s) + CO_2(g)$	2	3	1 (T or P)
$CuSO_4 \cdot 5H_2O(s)$, $CuSO_4 \cdot 3H_2O(s)$, $CuSO_4 \cdot H_2O(s)$, $CuSO_4(s)$, $H_2O(l)$	2	5	-1 (equilibrium impossible)

Systems are classified according to the phase rule as a *one-component system* (q.v.), a *two-component system* (q.v.) or a *three-component system* (q.v.).

See also Bo, F & C, F & J.

Phosphorus system

The phosphorus system is a *one-component system* (q.v.) exhibiting *mono-*

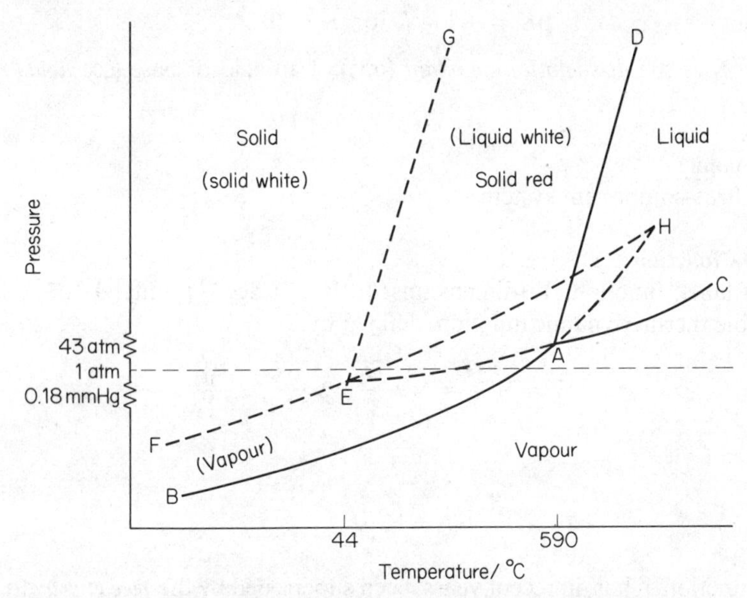

Figure P.8 Schematic diagram of the phosphorus system.

tropy (q.v.). Of the common forms of phosphorus, the red form is stable and the white metastable at all temperatures and pressures up to the transition point, H (figure P.8). The white variety is readily manufactured and the transition to the red is so slow that it can be stored almost indefinitely in the absence of air and light; transformation is accelerated by the addition of iodine.

In the simplified diagram, AB, AC and AD represent the two-phase equilibria for red phosphorus with a *triple point* (q.v.) at A (590 °C and 43 atm). Lines EF, EG and EA are the corresponding two-phase equilibria for the yellow variety with a triple point at E (44 °C and 0.18 mmHg). If curves BA and FE are extended, they meet at H, the triple point at which red and yellow phosphorus would be in equilibrium with the vapour. This point cannot be achieved experimentally, since both forms melt below this temperature. At 1 atm pressure, white phosphorus melts at 44.1 °C and boils at 280 °C (changing rapidly to the red variety above 250 °C); at 1 atm pressure, red phosphorus sublimes at 416 °C.

p*K*

pK is defined by the equation

pK

$$\text{p}K = -\log K \text{ or } K = 10^{-\text{p}K}$$

where K is the *dissociation constant* (q.v.) of an acid or base (see *acids and bases*).

Plait point
See Three-component system.

Planck function
The Planck function, Y (dimensions: m l^2 t^{-2} deg^{-1}; units: J K^{-1}), is a possible thermodynamic function, defined by

$$Y = -\frac{G}{T} = S - \frac{H}{T} = S - \frac{U}{T} - \frac{PV}{T}$$

whence

$$\mathrm{d}Y = \frac{H}{T^2}\,\mathrm{d}T - \frac{V}{T}\,\mathrm{d}P$$

The function Y has, in recent years, been superseded by the *free energy* (q.v.), G.

Polymorphism
Polymorphism is the occurrence of the same chemical substance in two or more different crystalline forms. Physical factors such as temperature, pressure and concentration, as well as the presence of foreign substances, help to modify the crystal habit. Interconversion of polymorphs occurs at a given temperature for a specified pressure; not all transitions are reversible. Examples:

$$\text{zinc blende (ZnS)} \xrightarrow{1024\ ^\circ\text{C}} \text{wurtzite (ZnS)}$$

$$\text{quartz } (SiO_2) \xrightarrow{870\ ^\circ\text{C}} \text{tridymite } (SiO_2) \xrightleftharpoons{1470\ ^\circ\text{C}} \text{cristobalite } (SiO_2)$$

Polymorphism in elements is called *allotropy* (q.v.).

Pressure
The pressure, P (dimensions: m l^{-1} t^{-2}; units: N m^{-2}; practical units: mmHg and atmosphere), of a gas is the force exerted by the gas per unit area. 1 atmosphere = 760 mmHg = 101 325 N m^{-2}. Thermodynamically,

$$P = f(V, T)$$

$$dP = \left(\frac{\partial P}{\partial V}\right)_T dV + \left(\frac{\partial P}{\partial T}\right)_V dT$$

Hence, for a system at constant pressure,

$$\left(\frac{\partial P}{\partial T}\right)_V = -\frac{(\partial V/\partial T)_P}{(\partial V/\partial P)_T}$$

The pressure is also related to the *partition function* (q.v.):

$$P = -\left(\frac{\partial A}{\partial V}\right)_{T,N} = kT\left(\frac{\partial \ln Q}{\partial V}\right)_{T,N}$$

Probability
See Entropy; Statistical thermodynamics.

Property
Properties are classified according to the manner in which they depend upon the size of the system. Extensive properties (or capacity factors), such as volume and mass, are proportional to the quantities of materials present in the system. Such properties are additive, i.e. the value of the property of the whole is the sum of the values of the constituent parts. U, H, S, G and A are all extensive properties. Intensive properties (or intensity factors) such as pressure, density and temperature, are independent of the amount of material present. Intensive properties are not additive.

All forms of energy can be considered as the product of these two types of property (see table P.5); the intensity factor of the energy always decreases in any spontaneous change.

Table P.5

Energy	Intensity factor	Capacity factor	Product
Mechanical	force (N)	distance (m)	N m = J
Mechanical (potential)	gh (m^2 s^{-2})	mass (kg)	kg m^2 s^{-2} = N m
Mechanical (kinetic)	velocity^2 (m^2 s^{-2})	mass (kg)	kg m^2 s^{-2} = N m
Mechanical	pressure (N m^{-2})	volume (m^3)	N m
Surface	surface tension (N m^{-1})	area (m^2)	N m
Electrical	potential diff. (V)	quantity of electricity (C)	V C = J
Thermal	temperature (K)	heat capacity or entropy (J K^{-1})	J
Chemical	chemical potential (J mol^{-1})	mole of reactant (mol)	J

Property

See also Colligative property.

R

Raoult's law

For an ideal mixture of two volatile components, at constant temperature, the partial *vapour pressure* (q.v.) of one component is proportional to its *mole fraction* (q.v.) in the liquid; the proportionality constant is the vapour pressure of the pure component, i.e.

$$p_A = p_A^{\ominus} x_A \quad \text{and} \quad p_B = p_B^{\ominus} x_B$$

If B is non-volatile, $p_B^{\ominus} = 0$; hence, the total pressure above the solution is given by

$$p = p_A = p_A^{\ominus} x_A$$

whence

$$x_B = \frac{p_A^{\ominus} - p}{p_A^{\ominus}}$$

This is the alternative statement of Raoult's law when only one component is volatile: the relative lowering of the vapour pressure of a volatile solvent (A) by a non-volatile solute (B) is equal to the mole fraction of the solute. In practice, the law only applies to dilute solutions of non-volatile solutes when x_A tends to 1 and to a limited number of mixtures of two volatile components over the whole range of concentrations; these are *ideal solutions*

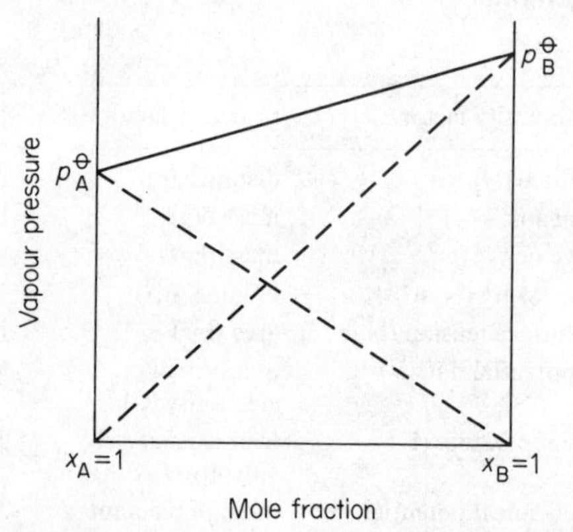

Figure R.1 Pressure–composition curve for an ideal solution.

(q.v.). The total vapour pressure above such an ideal binary liquid mixture is the sum of the partial vapour pressures (figure R.1):

$$P = p_A + p_B = (p_A^{\ominus} - p_B^{\ominus})\, x_A + p_B^{\ominus}$$

The composition of the vapour in equilibrium with an ideal solution is given by the application of Dalton's law:

$$\text{mole fraction of A in vapour} = y_A = \frac{p_A}{p_A + p_B} = \frac{p_A^{\ominus} x_A}{p_A^{\ominus} x_A + p_B^{\ominus} x_B}$$

if $p_A^{\ominus} = p_B^{\ominus}$, then $y_A = x_A$, and so the liquid and vapour have the same composition; while if $p_A^{\ominus} > p_B^{\ominus}$, then $y_A > x_A$, and the vapour is richer in the more volatile component.

Real solutions (q.v.) show marked deviations from Raoult's law, depending on the forces of attraction between the like and unlike molecules. On distillation, *binary liquid mixtures* (q.v.) may be separated into the pure components or one pure component and an azeotropic mixture (see *azeotrope*).

Reaction isochore
See van't Hoff isochore.

Reaction isotherm
See van't Hoff isotherm.

Real solutions
Real solutions show deviations from *Raoult's law* (q.v.) and *Henry's law* (q.v.) (compare *ideal solutions*). Although no solution is completely ideal, some solutions, e.g. benzene and toluene, obey these laws tolerably well over the

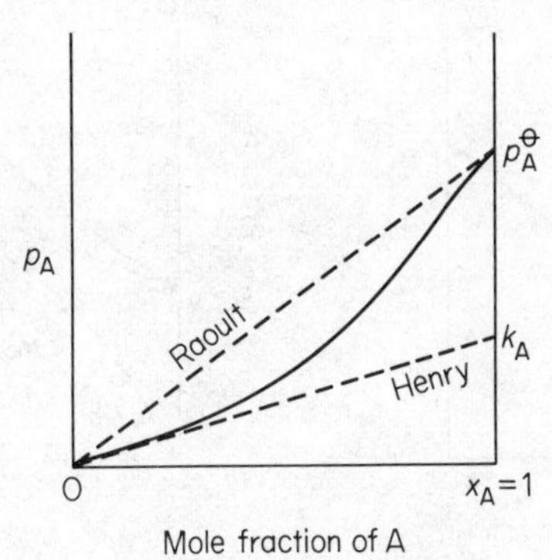

Figure R.2 Negative deviation.

whole concentration range, since the molecules are similar. When the components of a mixture are less similar, greater deviations from ideality are observed.

Negative deviations occur when the forces of attraction between unlike molecules are great; in consequence, the escaping tendencies of both types of molecule are reduced (figure R.2). As $x_A \rightarrow 1$, Raoult's law is valid; while as $x_A \rightarrow 0$, Henry's law is applicable, $p_A^\ominus > k_A$. Both components in a mixture show negative deviations, and if these are large enough, there will be a minimum in the v.p.–composition curve (figure R.3), giving rise to a maximum boiling point *azeotrope* (q.v.) (e.g. HCl–H_2O, chloroform–acetone). The large attraction between unlike molecules is in effect a chemical reaction, and mixing is accompanied by an evolution of heat and a contraction of volume. The association in solution tends to restrict the motion of the molecules and gives the system a lower *entropy* (q.v.) than in the ideal case (figure R.4).

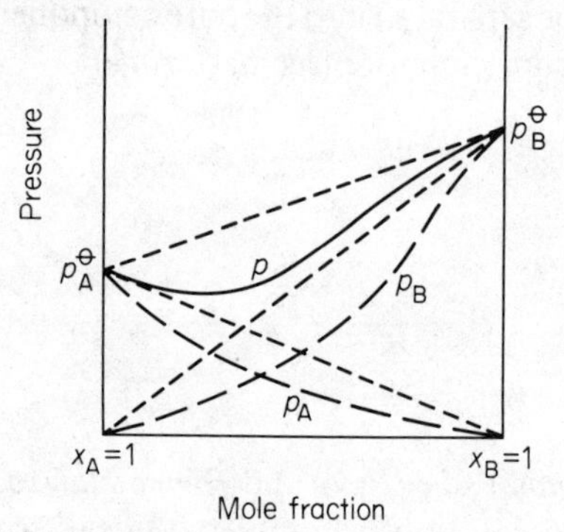

Figure R.3 System showing minimum in v.p.–composition curve.

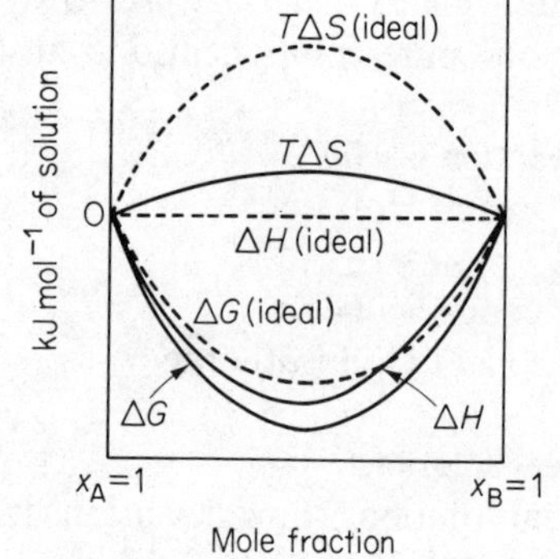

Figure R.4 Changes of thermodynamic functions on mixing.

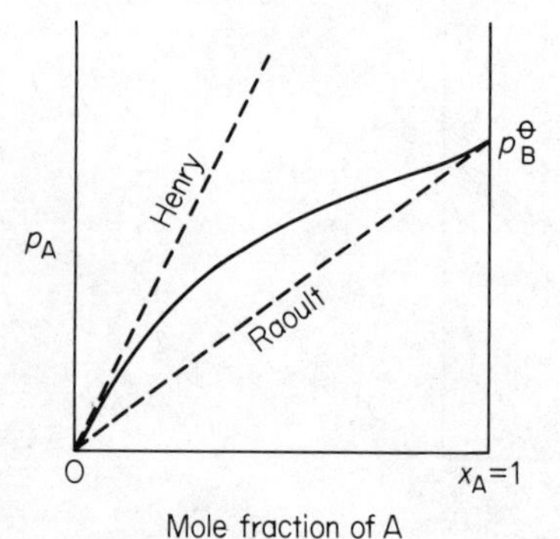

Figure R.5 Positive deviation.

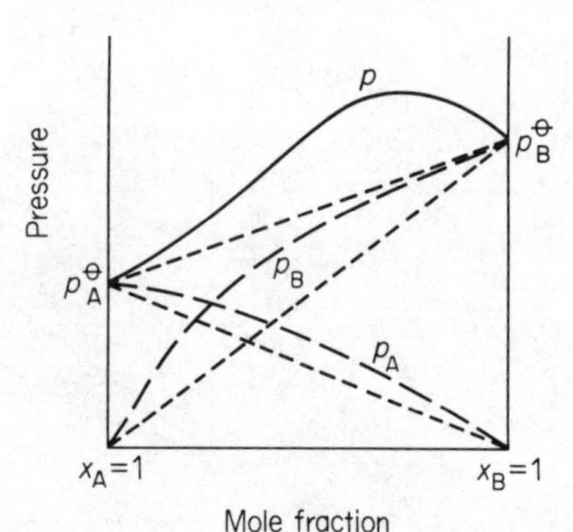

Figure R.6 System showing maximum in v.p.–composition curve.

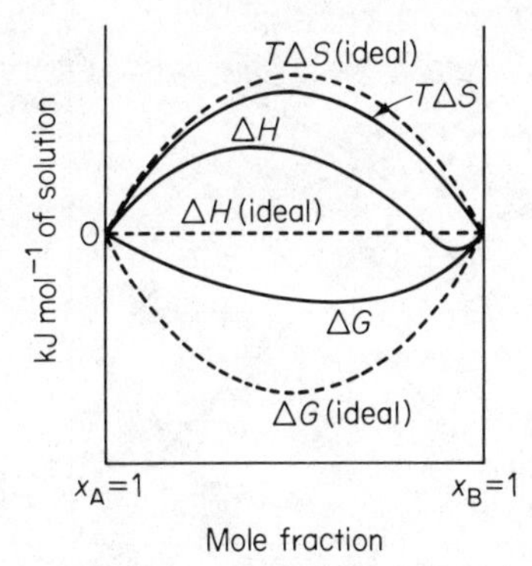

Figure R.7 Changes of thermodynamic functions on mixing.

Positive deviations occur when the forces of attraction (association) between like molecules are large; the molecules are thus less attracted into bulk solution and the escaping tendency for molecules of both types is increased (figure R.5). Again, as $x_A \rightarrow 1$, Raoult's law is valid; while as $x_A \rightarrow 0$, Henry's law is applicable, but now $k_A > p_A^\ominus$. Large deviations give rise to a maximum in the v.p.–composition curve (figure R.6) and, hence, a minimum boiling point azeotrope (e.g. ethanol–water, carbon tetrachloride–methanol). Mixing tends to break up some of the association and heat is absorbed to accomplish this. The entropy of mixing is less than than for the ideal case (figure R.7).

Refrigerator

Since every operation in a *Carnot cycle* (q.v.) is reversible, the cycle can be traversed in the reverse direction. In this case, an amount of heat, q_1, will be taken from the sink at the lower temperature and an amount of heat, q_2, put into the source; the amount of work done during the cycle is again given by the area enclosing the cycle, but it is now work done *on* the working substance. Working in reverse, the engine is acting as a heat pump or refrigerating machine. The effectiveness of the process or the coefficient of performance is

$$\frac{\text{Heat abstracted from the cold body}}{\text{Work done on the working substance}} = \frac{q_1}{q_2 - q_1}$$

Regular solutions

Regular solutions are solutions formed by two components in which the

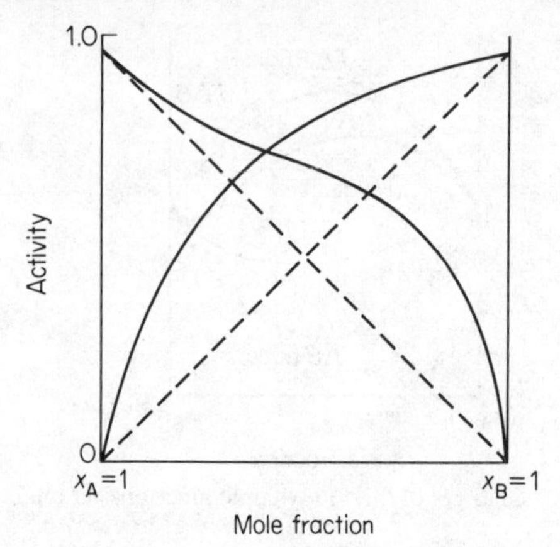

Figure R.8 Variation of activity with mole fraction for two components which form a regular solution.

entropy (q.v.) of mixing has the value expected for an *ideal solution* (q.v.), but for which the *enthalpy* (q.v.) of mixing is not zero; deviation of either component can be satisfactorily expressed by

$$\ln a_A = \ln x_A + B\, x_A^2$$
$$\ln a_B = \ln x_B + B\, x_B^2$$

from which it is apparent that the two activity–concentration curves should be symmetrical (figure R.8). In general, really symmetric curves are not obtained, since a larger power series is required to represent $\ln a_A$, etc.

Relative permittivity
See Dielectric constant.

Residual entropy
See Spectroscopic entropy.

Resonance energy
Some organic compounds, in which two or more valence structures are equally appropriate, have higher heats of atomisation than those calculated on the basis of the sum of the individual *bond energies* (q.v.). The extra stability is ascribed to resonance between the various structures, and the heat of atomisation, above that calculated for the most stable valence bond structure, is known as the resonance energy.

The heat of atomisation of benzene is obtained from the following data (all at 298 K):

$$C_6H_6(g) \longrightarrow 6C(\text{graphite}) + 3H_2(g) \quad \Delta H = -82.9 \text{ kJ}$$
$$6C(\text{graphite}) \longrightarrow 6C(g) \quad \Delta H = 6 \times 715.5 = 4293.0 \text{ kJ}$$
$$3H_2(g) \longrightarrow 6H(g) \quad \Delta H = 3 \times 435.1 = 1305.3 \text{ kJ}$$

Hence,

$$C_6H_6(g) \longrightarrow 6C(g) + 6H(g) \quad \Delta H = 5515.4 \text{ kJ}$$

From bond energy data, the heat of atomisation of the Kekulé structure is given by

$$\Delta H = 3\varepsilon(\text{C—C}) + 3\varepsilon(\text{C=C}) + 6\varepsilon(\text{C—H}) = 5346 \text{ kJ}$$

The resonance energy, 169 kJ, is the difference between the calculated and observed heats of atomisation. Thus more heat is evolved in the formation of benzene from gaseous atoms than would be expected on the basis of Kekulé structures.

Reversible process

A thermodynamically reversible process is one which may be followed in either direction and in which, at any point during the process, the direction may be reversed by an infinitesimally small change in a variable, e.g. temperature or pressure. A reversible process is one which, in the limit, proceeds infinitely slowly through a sequence of equilibrium states. For the process to occur, there must be a slight difference between the driving (external) force and the opposing (internal) force. The advantage of reversibility is

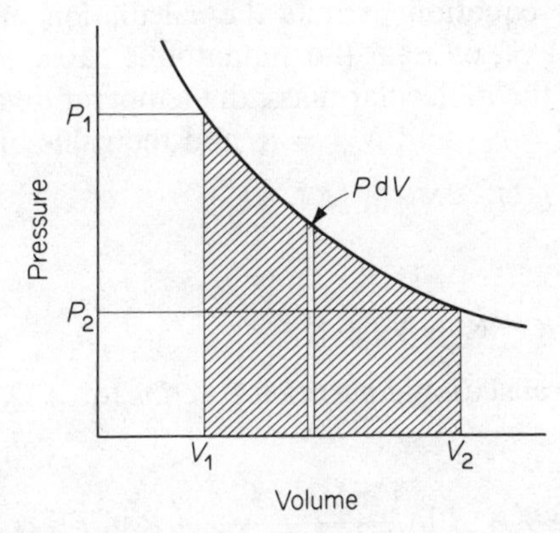

Figure R.9 Pressure–volume isothermal for the reversible expansion of an ideal gas.

that it introduces the idea of continuity of any change, and makes it possible to draw a graph to represent the change (since each state is in a condition of equilibrium), and, hence, the *state functions* (q.v.) are completely fixed. In the isothermal expansion of n mole of an ideal gas, the graph of P against V represents the expansion (figure R.9) and the *work* (q.v.) done is given by

$$w = n \int_{V_1}^{V_2} P \, \mathrm{d}V = \text{Area under the } P\text{–}V \text{ curve} = nRT \ln V_2/V_1$$

In an *irreversible process* (q.v.), no such graphical representation can be made, since no unique values can be assigned to all the variables. When a process is carried out reversibly, the maximum amount of work is obtained.

The nearest approximation to a reversible process in the laboratory is an electric cell, the potential of which is just balanced by an opposing potential, E(outside), as in a potentiometer circuit. When E(cell) is infinitesimally greater than E(outside), the cell acts as a *galvanic cell*†, the cell reaction proceeds infinitely slowly and an infinitesimal current flows; when E(outside) is slightly greater than E(cell), the cell becomes an *electrolytic cell*†, current is reversed in the cell and the cell reaction proceeds in the opposite direction. When E(cell) = E(outside), no current flows and no work is done.

See also Adiabatic process; Isobaric process; Isochoric process; Isothermal process; and D & J, K, Was.

S

Sackur–Tetrode equation

The Sackur–Tetrode equation permits the calculation of the total translational *entropy* (q.v.), S_t, of a gas (for monatomic gases, the total entropy) from a knowledge of the molecular mass, the temperature and the pressure. For 1 mole of gas, $N = N_A$ and $N_A k = R$, and the value of S_t given by

$$S_t = Nk \ln Q + U/T - Nk \ln N + Nk \qquad \text{(P.7a)}$$

becomes

$$S_t = R \ln Q + \tfrac{3}{2}R - R \ln N_A + R$$

Substituting for the translational *partition function* (q.v.), Q, and rearranging ($M = mN_A$) gives

$$S_t = \tfrac{5}{2}R + \tfrac{5}{2}R \ln T + R \ln \frac{(2\pi M)^{3/2} \, k^{5/2}}{h^3 \, N_A^{3/2}} - R \ln P$$

Substituting the values of the constants

$$S_t = 28.72 \log M + 47.87 \log T - 19.14 \log P + 172.79$$

at $T = 298.15$ K,

$$S_t = 28.72 \log M + 291.29 - 19.14 \log (P/N\ \mathrm{m}^{-2})$$

or, when $P = 101\ 325\ \mathrm{N\ m^{-2}}$, the value of $S_t/\mathrm{J\ K^{-1}\ mol^{-1}}$ is given by

$$S_t = 28.72 \log (M/\mathrm{kg\ mol^{-1}}) + 195.48$$

Thus, at 298.15 K and at $P = 101\ 325\ \mathrm{N\ m^{-2}}$, the following values of S_t are obtained:

Gas	Molecular mass/kg $\mathrm{mol^{-1}}$	$S_t/\mathrm{J\ K^{-1}\ mol^{-1}}$
Ar	0.039 948	155.30
HCl	0.036 47	154.18
N_2O	0.044 013	156.56

The translational entropy of a gas at normal temperatures provides the greatest contribution to the total entropy of a gas (see *spectroscopic entropy*). The value of S_t for argon agrees well with the third law value of 154.7 J $\mathrm{K^{-1}\ mol^{-1}}$.

Salt hydrates

Salt hydrates are examples of *two-component condensed systems* (q.v.) in which the components are a salt and water. Most salt hydrates decompose

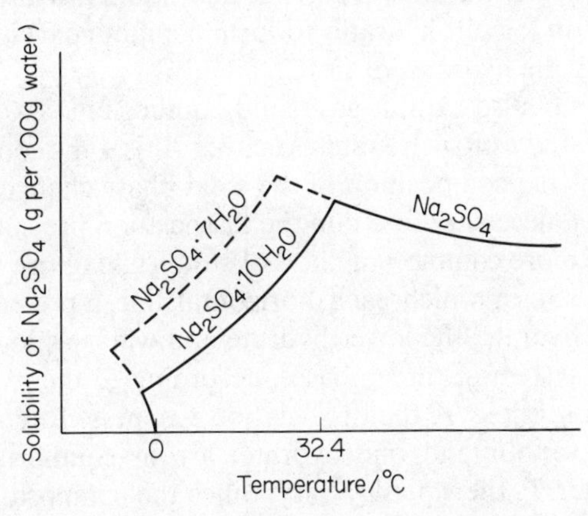

Figure S.1 Solubility curve of sodium sulphate.

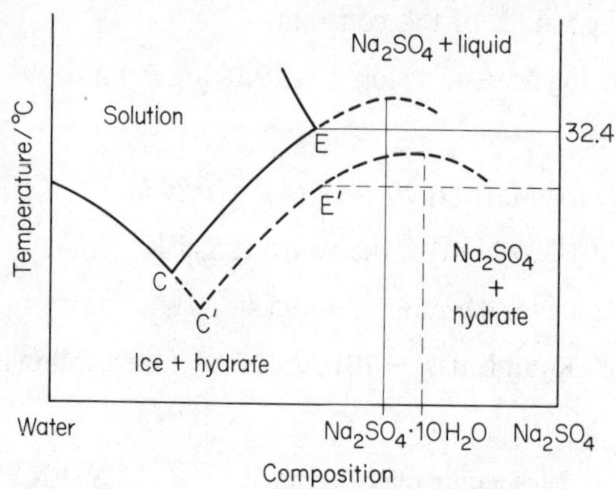

Figure S.2 Phase diagram for the sodium sulphate–water system.

below their melting point into water or a solution, and the anhydrous salt or a lower hydrate. Such systems are common examples of two-component condensed systems forming a compound with an incongruent m.p.

The axes of the solubility curve for sodium sulphate (figure S.1), with the breaks caused by the dehydration of the hydrates, may be changed to give the normal phase diagram (figure S.2). The curves CC′ and C′E′ give the equilibrium states for $Na_2SO_4 \cdot 7H_2O$, which may be deposited if a solution is cooled rapidly below CE so that there is insufficient time for the decahydrate to crystallise out. C′E′ represents the metastable equilibrium between liquid and heptahydrate, and E is the incongruent m.p. where the decahydrate, the anhydrous salt and liquid are in equilibrium.

It is less common for salt hydrates to form compounds with a congruent melting point (e.g. the hydrates of Fe_2Cl_6).

Salt hydrates may also exhibit gas–solid equilibrium. Two methods are available for the study of such systems: either T is kept constant and the P is determined as the composition of the solid phase changes, or P is kept constant and the T necessary to attain the dissociation pressure is observed. The former is the more common method of study, giving rise to the stepped isotherm (figure S.3), in which each horizontal line represents the v.p. at which the higher hydrate, the lower hydrate and water vapour are in equilibrium. For this system of three phases, according to the *phase rule* (q.v.), $f = 1$; hence, at a selected T, the dissociation v.p. must be constant. When two phases (the vapour and one hydrate) are in equilibrium, $f = 2$, so that, at a particular T, the v.p. may vary. Since the composition of hydrate is fixed, the vertical line shows this variation of v.p. In the isothermal

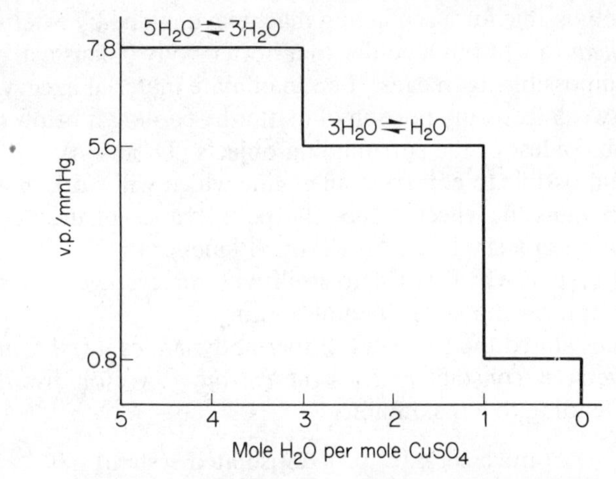

Figure S.3 Isothermal (25 °C) dehydration of $CuSO_4 \cdot 5H_2O$.

dehydration (under reduced pressure) of $CuSO_4 \cdot 5H_2O$, the equilibrium pressure (7.8 mmHg) remains constant while the transformation to the trihydrate occurs; after this is completed, the v.p. falls rapidly to 5.6 mmHg, where it remains constant until the transformation to the monohydrate has been completed. Thereafter the pressure falls to 0.8 mmHg until eventually the anhydrous salt remains.

The equilibrium between salts and ammonia gas (e.g. $AgCl \cdot NH_3$) is analogous to that between a salt and water vapour.

See also Deliquescence; Efflorescence.

Second law of thermodynamics

The second law of thermodynamics is a general statement of natural tendencies, distinguishing changes which occur of their own accord and those which do not occur under given circumstances. In contrast, the *first law of thermodynamics* (q.v.), a statement of the equivalence of different forms of energy, gives no guidance as to the conditions under which these changes may occur. The second law gives no information about the rate of *spontaneous processes* (q.v.). The second law has been stated in many different ways.

(1) Spontaneous processes are those which when carried out under proper conditions can be made to do *work* (q.v.) (if carried out reversibly, the maximum amount of work will be obtained; in the natural irreversible way, the maximum amount of work is never obtained).

(2) It is impossible for a self-acting machine, unaided by external agency, to transfer *heat* (q.v.) from a colder to a hotter body (Clausius).

(3) It is impossible, by means of an inanimate material agency, to derive mechanical work from any portion of matter by cooling it below the temperature of the coldest of the surrounding objects (Thomson).

(4) It is impossible to construct an engine which will work in a complete cycle and produce no effect except the performance of mechanical work and the cooling of a single heat reservoir (Planck).

(5) Every system which is left to itself will, on average, change in such a way as to approach a state of equilibrium.

(6) Clausius stated the two laws of thermodynamics: (a) the *energy* (q.v.) of the Universe is constant and (b) the *entropy* (q.v.) (or disorder) of the Universe is tending to a maximum:

$$\Delta S(\text{universe}) \geq 0 \quad \text{or} \quad \Delta S(\text{isolated system}) \geq 0$$

Since a system whose U, V and mass remain constant is in effect an isolated system, it follows that

$$\Delta S_{U,V,m} \geq 0$$

Experience shows that some energy transformations are possible and may occur spontaneously, while energy transformations in the reverse direction are impossible. Thus water (electricity) flows from a higher level (potential) to a lower, but never in the reverse direction. When an object slides down an inclined plane, heat is produced by friction; this heat can never be completely converted back into energy and is never sufficient to raise the object back to its original position. Similarly, the chemical energy produced when coal is completely burnt in oxygen can be completely converted into heat, but the products of combustion and the heat cannot be converted back to coal and oxygen.

See also Carnot cycle; Carnot theorem; and Be.

Sign convention in thermodynamics
See State function.

Solid solutions
Solid solutions are formed only when one solid can dissolve in another to form a homogeneous mixture. In many ways, they are analogous to liquid solutions. If the two components are mixed, they will diffuse into one another and form a solution. Many alloys are solid solutions.

See also Two-component condensed systems.

Solubility product

For a uni-univalent sparingly soluble salt MA, the solubility product, or, activity solubility product, K_s (units: mol$^{\nu}$ kg$^{-\nu}$), is defined by

$$K_s = a(M^+)\,a(A^-) \quad \text{since} \quad a(MA, s) = 1$$

or, for a salt which dissociates into ν^+ positive ions of charge z^+ and ν^- negative ions of charge z^-,

$$K_s = a_{z+}^{\nu+}\ a_{z-}^{\nu-}$$

These equations apply to any saturated solution, but are only of any practical interest in the case of sparingly soluble salts. For very dilute solutions, the *activity coefficients* (q.v.) can be taken as unity and, hence,

$$K_s = m(M^+)\,m(A^-)$$

If the salt is completely dissociated $m_+ = m_- = K_s^{1/2}$. So long as pure solid MA is in equilibrium with the dissolved solute MA, any modification which causes a change in the solubility of MA ($m_\pm$), such as the addition of an electrolyte, must be due to a change in the activity coefficient, $\gamma_\pm$, since, at constant T and P, K_s is constant:

$$K_s = a(M^+)\,a(A^-) = a_\pm^2 = m_\pm^2\,\gamma_\pm^2 \tag{S.1}$$

If the added electrolyte contains an ion (A^-) in common with MA, the increase in ionic strength causes $\gamma_\pm$ to decrease, but this is outweighed by the increase in $m(A^-)$ due to the added A^-. Thus, for K_s to remain constant, $m_\pm$ must decrease; this is reflected in a decreased solubility, known as the 'common ion effect' (figure S.4).

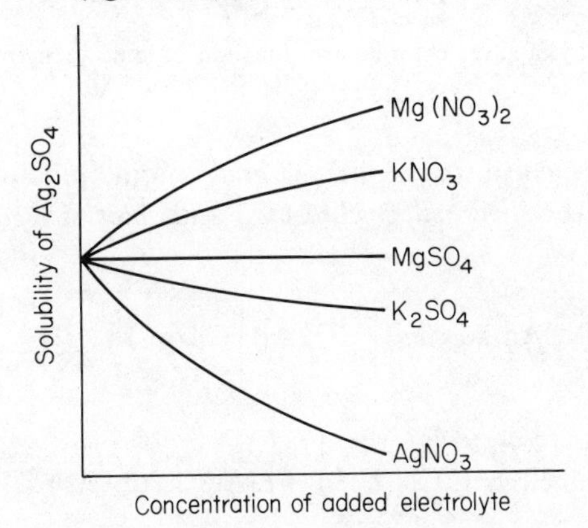

Figure S.4 Influence of various salts on the solubility of silver sulphate.

Solubility product

If the added electrolyte contains no common ion, $\gamma_\pm$ decreases and, hence, $m_\pm$ and thus the solubility increase to maintain K_s constant. At very high electrolyte concentrations, a point is reached when $\gamma_\pm$ increases and $m_\pm$ starts to decrease.

Determination of K_s

From solubility measurements in the presence of added electrolytes Equation (S.1) can be written in the form

$$\log \gamma_\pm = \log K_s^{1/2} - \log m_\pm$$

If $m_\pm$ is measured in the presence of known amounts of added electrolyte, the graph of $\log m_\pm$ against $I^{1/2}$ is linear and of intercept $\log K_s^{1/2}$, whence K_s and, hence, the activity coefficient at any concentration can be obtained (figure S.5).

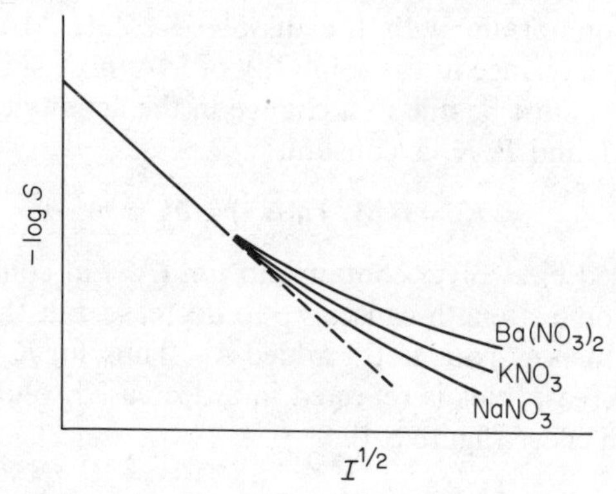

Figure S.5 Log solubility of silver chloride as a function of ionic strength of the supporting electrolyte.

From e.m.f. measurements using a cell† *without a* liquid junction†. For AgCl, the cell containing the *silver, silver chloride*† and *chlorine electrodes*† would be suitable:

$$\ominus \quad \text{Ag, AgCl(s)} \,\big|\, \text{HCl(aq)} \,\big|\, \text{Cl}_2\text{(g), Pt} \quad \oplus$$

for which

$$E(\text{cell}) = E(\text{Cl}_2, \text{Cl}^-) - E(\text{AgCl, Ag, Cl}^-)$$

whence

$$\log K_s = \{E^{\ominus}(Cl_2, Cl^-) - E^{\ominus}(Ag^+, Ag) - E(\text{cell})\}\frac{F}{2.303\,RT}$$

See also D & J, J & P, K.

Spectroscopic entropy

The spectroscopic entropy of a gas can be obtained from a knowledge of the translational, rotational and vibrational *partition functions* (q.v.). The translational entropy, arising from translational motions of the centres of mass of the molecules, is calculated from the *Sackur–Tetrode equation* (q.v.), which, at 298.15 K and $P = 101\,325$ N m^{-2}, is

$$S_t = 28.72 \log M + 195.48 \text{ J K}^{-1} \text{ mol}^{-1}$$

The rotational entropy, due to unhindered rotation of molecules as rigid bodies about the centre of mass, for 1 mole of gas is given by

$$\begin{aligned} S_r &= N_A k \ln Q_r + N_A kT\left(\frac{\partial \ln Q_r}{\partial T}\right)_V \\ &= R \ln \frac{8\pi^2 kTI}{h^2\sigma} + R \\ &= 19.14 \log TI + 877.5 - 19.14 \log \sigma \end{aligned}$$

(where I is the moment of inertia and σ the *symmetry number*, q.v.). The vibrational entropy, arising from the thermally excited bond-bending and bond-stretching vibrations for 1 mole of gas, is given by

$$\begin{aligned} S_v &= N_A k \ln Q_v + N_A kT\left(\frac{\partial \ln Q_v}{\partial T}\right)_V \\ &= R\left[\frac{h\nu/kT}{\{\exp(h\nu/kT) - 1\}} - \ln\{1 - \exp(-h\nu/kT)\}\right] \end{aligned}$$

(where ν is the characteristic frequency of vibration).

Table S.1 shows the relative magnitudes of the contributions from translation, rotation and vibration to the total entropy for several gases (assumed to be ideal). Two important points are apparent; firstly, the dominant contribution is the entropy associated with translational motion; and, secondly, the contribution from the vibrational motion is negligibly small (this is due to the fact that, at 298 K, kT is only equivalent to a frequency of 6×10^{12} s^{-1}, and, in general, all the vibrational frequencies are larger than this).

Table S.1 also gives a comparison between the spectroscopic and the

Table S.1. Spectroscopic entropy values (/J K^{-1} mol^{-1}) at $T = 298.15$ K and $P = 101\ 325$ N m^{-2}

	Ar	HCl	N_2O	CO	H_2O(g)
Molar mass/kg mol^{-1}	0.039 948	0.036 465	0.044 013	0.028 01	0.018 015
Moment of inertia (10^{47} I/kg m^2)		2.71	66.9	14.48	1.024 1.921 2.947
Symmetry number		1	1	1	2
Vibrational freq. (10^{-13} ν/s^{-1})		8.654	1.765($g = 2$) 3.854 6.666	6.5	4.773 10.95 11.27
S_t	155.30	154.18	156.53	150.88	145.39
S_r	0	33.19	60.30	47.4	43.74
S_v	0	10^{-5}	3.96	0	0
$S_{spec,}$	155.30	187.37	220.79	198.28	189.13
$S_{thermal}$	154.70	186.60	215.20	193.36	185.20
$S_{spec,} - S_{thermal}$			5.59	4.92	3.93

thermal *entropy* (q.v.) values; for Ar and HCl, the difference is negligible. The closeness of these values, an agreement which also occurs for a wide range of diatomic molecules, is evidence of both the experimental skill in determining $S_{thermal}$ by calorimetric methods and the accuracy of modern knowledge of the nature of simple molecules in the gas phase, on which statistical calculations are based. The discrepancies for CO, N_2O and H_2O are therefore due to a real departure of the probability, W, from the value of unity and, hence, of S_0 from the value of zero at 0 K. These differences are, in fact, the values of the residual entropy at 0 K; i.e.

$$S_{spec.} = S_{thermal} + k \ln W$$

This means that, for these molecules, at 0 K the perfectly ordered crystal structure is not attained. CO is a linear asymmetric molecule; the disorder arises from the fact that the ends of the molecule are similar and the crystal might form, not with the perfect order (e.g. CO—CO—CO—CO—) but possibly with a disordered pattern (e.g. CO—CO—OC—CO—). A crystal formed in this way has insufficient energy at these low temperatures for the molecules to be reoriented. The entropy of such a crystal, on probability grounds, might be expected to be $k \ln (2)^{N_A} = 5.74$ J K^{-1} mol^{-1}, a value close to the discrepancy shown by these gases.

A similar explanation has been advanced to explain the result $k \ln W = 3.93$ for water. In detail, the theory for water gives $W = 3/2$; hence, $k \ln (3/2)^{N_A} = 3.4$ J K^{-1} mol^{-1}, in remarkable agreement with the experimental value.

See also Boltzmann equation; Third law of thermodynamics; and K, N, Wi.

Spontaneous process

A spontaneous process, or natural process, is one which is capable of doing *work* (q.v.). All spontaneous processes are irreversible, i.e. they cannot be reversed merely by changing parameters by infinitesimally small amounts. Examples of such processes include nuclear fission, the flow of heat from a hotter to a colder body, the expansion of a gas and the solution of a solute in a solvent. Such changes are accompanied by a decrease in the *free energy* (q.v.), $\Delta G < 0$, and an increase in the *entropy* (q.v.) or disorder of the system, $\Delta S > 0$.

See also Second law of thermodynamics.

Standard states

Standard state of a solid is the most stable form of the pure solid at a pressure of 101 325 N m^{-2} and the specified temperature; activity of solid = 1.

Standard state of a liquid is the most stable form of the liquid at a pressure of 101 325 N m^{-2} and the specified temperature; activity of a liquid = 1.

Standard state of an ideal gas is the pure gas at a pressure of 101 325 N m^{-2} and the specified temperature; fugacity = 101 325 N m^{-2}.

Standard state of a real gas is the pure gas at the specified temperature and a thermodynamic pressure (fugacity) of 101 325 N m^{-2}, where the enthalpy is that of a real gas at zero pressure. Little error is involved if the standard pressure of a real gas is taken as 101 325 N m^{-2}.

Standard temperature is 298.15 K.

Standard thermodynamic functions

See Enthalpy; Entropy; Free energy; Internal energy.

State function

When the state of any system is changed, the change of any state function, or state variable, depends only on the value of the function in the initial and final states; it is independent of the path along which the change has occurred. For a function X,

$$\Delta X = X_2 - X_1 = \int_{X_1}^{X_2} \mathrm{d}X$$

Increase in value of X = Final value of X − Initial value of X

Standard convention is that ΔX always represents an increase in the value of X

State functions, e.g. T, P, V, S, U, H, G, A and density, are not necessarily independent of one another. In a specified amount of gas, V and the density are determined by P and T; the relationship between P, V and T for a given amount of substance is known as an equation of state; this depends on the nature of the substance. *Heat* (q.v.) and *work* (q.v.) are *not* state functions. Once the state of a system is specified by the values of a few state functions, the values of all the others are fixed.

See also Exact differential.

State variable
See State function.

Statistical thermodynamics
In thermodynamics, the molecular structure of matter is ignored and the transformation of energy from one form to another and its availability for doing work are considered without reference to the detailed structure of the system in which the changes take place. In molecular theory, matter is visualised as composed of very large numbers of particles in continual and random motion. The individual particles are subject to the laws of quantum mechanics, which are often indistinguishable from the laws of classical dynamics. Statistical mechanics provides the bridge between the dynamical properties of the particles and the thermodynamic properties of the entire assembly or system of particles.

Since very large numbers of particles are considered, statistical methods are used to calculate what fraction of the total number of particles have an assigned energy, and also to express the principal thermodynamic quantities of the assembly as functions of the energies of the particles. Since energy is quantised, the particles in an assembly can have certain definite levels of energy only. Suppose that there is a specified distribution of the particles among different energy levels such that n_1 have energy ε_1; n_2 have energy ε_2; n_3 have energy ε_3, etc. ($n_1, n_2, n_3 \ldots$ are known as distribution numbers). Then the probability or number of ways in which this arrangement can be made is given by

$$\frac{N!}{n_1!\, n_2!\, n_3! \ldots n_r!} = \frac{N!}{\prod_r n_r!}$$

The term 'probability' is used since it is assumed that all the ways in which the N particles can be assigned to the different energy levels are equally probable. This, the fundamental assumption of 'molecular chaos', means that there is no law directing particles to any one energy level in preference to any other, and that the interchange of particles among the different levels is completely random.

The total number, W, of different states of the assembly is given by the sum of expressions of the above type for all possible sets of distribution numbers; i.e.

$$W = \sum_{\substack{\text{all possible}\\ \text{sets of values}\\ \text{of } n_1, n_2 \ldots n_r}} \frac{N!}{\prod_r n_r!} \tag{S.2}$$

Statistical probability is also concerned with the arrangement of particles in space; thus the probability of finding a molecule in any given small volume element of gas increases with the volume of the element. When the system is more ordered, there are fewer ways of arranging the particles spatially, and, in a perfect crystal, there is only one way of arranging the particles. Left to themselves, the particles in a gas will tend to become very disordered with respect to position, orientation and energy, and the distribution will rapidly approach that which can be achieved in the greatest number of ways, i.e. W increases and is maximal at equilibrium, and, hence, so also is S, to which it is related by the *Boltzmann equation* (q.v.):

$$S = k \ln W \tag{S.3}$$

If two assemblies of probability W_A and W_B and entropy S_A and S_B, respectively, are combined, then the probability of the combined assembly is $W_A \times W_B$ and the total entropy is $S_A + S_B$. In a perfect crystal, $W = 1$ and $S = 0$.

In evaluating W for the assembly, there are two restrictions on the permissible values of $n_1, n_2, n_3 \ldots$, namely that the total number of the particles and the total energy are constant; expressed in equations:

$$n_1 + n_2 + n_3 + \ldots + n_r = \sum_r n_r = N \tag{S.4}$$

and

$$n_1 \varepsilon_1 + n_2 \varepsilon_2 + n_3 \varepsilon_3 + \ldots + n_r \varepsilon_r = \sum_r n_r \varepsilon_r = U \tag{S.5}$$

Thus

$$W = \sum_{\left(\substack{\sum_r n_r = N \\ \sum_r n_r \varepsilon_r = U}\right)} \frac{N!}{\prod_r n_r!} \tag{S.6}$$

This sum can only be evaluated when N is very large; under these conditions, only the largest term in the sum makes any effective contribution to W. Substituting equation (S.6) in equation (S.3) gives:

$$S = k \ln W = k \ln t_{\max} = kN \ln N! - k\sum_r \ln n_r!$$

Introducing Stirling's approximation for large values of N, i.e.

$$\ln N! = N \ln N - N$$

gives

$$\begin{aligned} S &= k[N \ln N - N - \sum_r (n_r \ln n_r - n_r)] \\ &= k[N \ln N - \sum_r n_r \ln n_r] \end{aligned} \tag{S.7}$$

The condition for a maximum is that

$$\partial S = \sum \mathrm{d}S_r = 0$$

for all possible variations of n_r. Differentiating equation (S.7) with respect to n_1 gives

$$\mathrm{d}S/\mathrm{d}n_1 = -k(\ln n_1 + 1)$$

and, similarly,

$$\mathrm{d}S/\mathrm{d}n_2 = -k(\ln n_2 + 1)$$

Since k cannot be zero (equation S.3), the condition for maximum S is

$$\sum_r (\ln n_r + 1)\,\mathrm{d}n_r = 0 \tag{S.8}$$

The two subsidiary conditions of constancy in the total number of particles (equation S.4) and in the total energy of the system (equation S.5) may be expressed:

$$\partial N = \sum \mathrm{d}n_r \quad = 0 \tag{S.9}$$

$$\partial U = \sum \varepsilon_r \mathrm{d}n_r = 0 \tag{S.10}$$

Lagrange's method of undetermined multipliers allows equations (S.8)–(S.10) to be satisfied simultaneously by the use of two adjustable constants α and β.

In this way,

$$\sum_r (\ln n_r + 1 + \alpha + \beta \varepsilon_r)\, dn_r = 0 \tag{S.11}$$

This equation is true for arbitrary variations, dn_r; hence, in general, all terms of the kind in the bracket (equation S.11) must separately equal zero, i.e.

$$\ln n_r + 1 + \alpha + \beta \varepsilon_r = 0$$

or

$$n_r = K \exp(-\beta \varepsilon_r) \tag{S.12}$$

where $K = \exp(-\alpha - 1)$. The total number of particles in the system is given by

$$N = \sum_r n_r = K \sum_r \exp(-\beta \varepsilon_r) \tag{S.13}$$

Eliminating K between equations (S.12) and (S.13) gives

$$n_r = \frac{N \exp(-\beta \varepsilon_r)}{\sum_r \exp(-\beta \varepsilon_r)} \tag{S.14}$$

Taking logarithms of equation (S.14) and multiplying through by n_r gives

$$n_r \ln n_r = n_r \ln N - n_r \beta \varepsilon_r - n_r \ln \{\sum_r \exp(-\beta \varepsilon_r)\} \tag{S.15}$$

The addition of all equations of this type gives

$$\sum_r n_r \ln n_r = N \ln N - \beta U - N \ln \{\sum_r \exp(-\beta \varepsilon_r)\} \tag{S.16}$$

Combining equations (S.7) and (S.16) gives

$$S/k = \beta U + N \ln \{\sum_r \exp(-\beta \varepsilon_r)\} \tag{S.17}$$

Thus

$$\left(\frac{\partial U}{\partial S}\right)_V = \frac{1}{k\beta}$$

Since the thermodynamic temperature is defined as

$$T = \left(\frac{\partial U}{\partial S}\right)_V$$

it follows that $\beta = 1/kT$, whence, according to the Maxwell–Boltzmann statistics, the number of particles with energy ε_r is obtained by rewriting equation (S.14) thus:

$$n_r = \frac{N \exp(-\varepsilon_r/kT)}{\sum_{r=0}^{r=\infty} \exp(-\varepsilon_r/kT)} = \frac{N \exp(-\varepsilon_r/kT)}{Q} \tag{S.18}$$

where Q is the *partition function* (q.v.).

In fact, there are frequently g_r energy states with nearly equal energy ε_r; these may be regarded as a single degenerate state of statistical weight g_r and thus W_{max} (c.f. equation S.2) becomes:

$$W_{max} = \frac{N! \prod_r (g_r)^{n_r}}{\prod_r n_r!}$$

and, according to the Bose–Einstein statistics for degenerate states, equation (S.18) becomes

$$n_r = \frac{N\, g_r \exp(-\varepsilon_r/kT)}{Q}$$

The entropy of the system may thus be obtained from equation (S.17) as

$$S = UT + Nk \ln Q$$

The relationships between other thermodynamic functions and the partition function follow from this.

See also Kn, N, Ru.

Steam distillation

Steam distillation is the process of cleaning up and concentrating organic materials of high molecular mass by passing steam into a heated mixture of the organic compound and water. An advantage of steam distillation is that the b.p. is reduced below 100 °C; many organic compounds boil at such high temperatures that they decompose. Both the organic compound and water, as *immiscible liquids* (q.v.), contribute separately to the total vapour pressure, and the amounts of organic compound and water collected in the distillate are proportional to their v.p. and relative molecular mass:

$$\frac{\text{Weight of organic compound in distillate}}{\text{Weight of water in distillate}} = \frac{p(\text{org}) \times M_r(\text{org})}{p(\text{H}_2\text{O}) \times M_r(\text{H}_2\text{O})}$$

thus best yields are obtained when p(org) and M_r (org) are large, but since such a compound has a high b.p., its v.p. is usually small.

Consider the steam distillation of nitrobenzene mixed with water containing about 1% nitrobenzene. This mixture will boil at 99.25 °C (figure S.6);

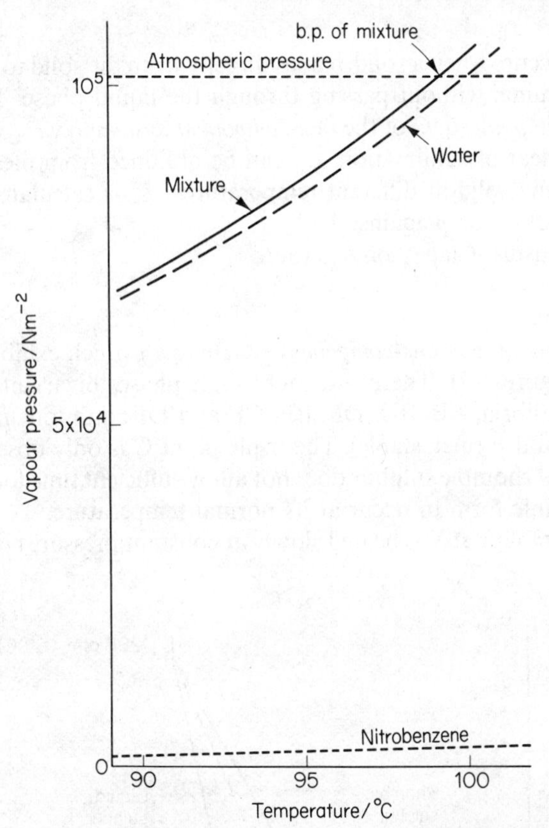

Figure S.6 Vapour pressure–temperature curves for the nitrobenzene–water mixture.

at this temperature, the v.p.s of water and nitrobenzene are 98 791.60 and 2 699.77 N m^{-2}, respectively. Thus

$$\frac{\text{Weight of nitrobenzene in distillate}}{\text{Weight of water in distillate}} = \frac{2\,699.77 \times 123}{98\,791.6 \times 18} = \frac{0.1865}{1}$$

i.e. there is about 15% by weight of nitrobenzene in the distillate. This represents a great concentration, apart from any purification.

Steam distillation is less efficient, although often used, if the liquids are partially miscible at the temperature of distillation.

The principle of steam distillation can be used to determine the v.p. of the pure organic liquid of known molecular mass from a measurement of the relative weights of the two phases in the condensed distillate.

Sublimation

Sublimation occurs when a solid passes directly from the solid to the vapour phase on warming, without passing through the liquid phase. This occurs below the *triple point* (q.v.) of the *one-component system* (q.v.).

The molar heat of sublimation, L_s, can be obtained from measurements of the v.p. of the solid at different temperatures; L_s is calculated from the slope of the plot of log p against T^{-1}.

See also Clausius–Clapeyron equation.

Sulphur system

The sulphur system is a *one-component system* (q.v.) which exhibits *enantiotropy* (q.v.) (figure S.7). There are four single-phase, bivariant areas; six two-phase equilibria, AB, BD, DE, BF, CF and DF; and four *triple points* (q.v.) B, C, D and F (metastable). The triple point C is only observed when rapid heating of rhombic sulphur does not allow sufficient time for transition to the monoclinic form to occur at its normal temperature.

If rhombic sulphur at V is heated slowly at constant pressure (760 mmHg),

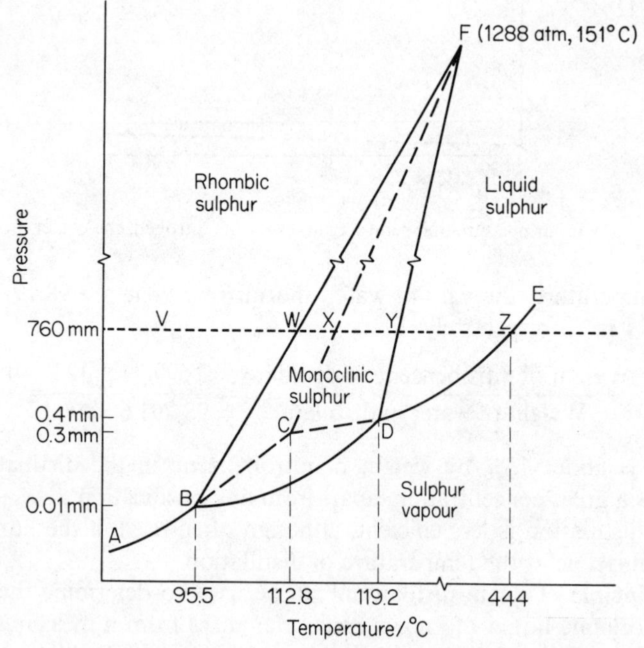

Figure S.7 Phase diagram for the sulphur system (not to scale).

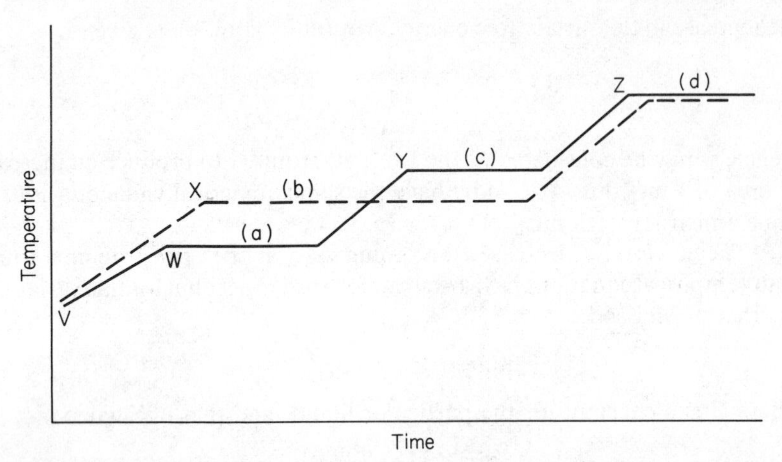

Figure S.8 Temperature–time curves for the uniform heating of sulphur at constant pressure (1 atm). (a) Temperature constant while rhombic S is transformed to monoclinic S. (b) Temperature constant while rhombic S melts. (c) Temperature constant while monoclinic S melts. (d) Temperature constant while liquid S boils.

it will be converted into the monoclinic form at W, melt at Y and boil at Z (figure S.8). Rapid heating along the same line will result in rhombic sulphur melting at X and boiling at Z.

Liquid sulphur exists in two forms, S_λ and S_μ.

Surface energy
See Surface tension.

Surface tension
The surface tension, γ (dimensions: m t^{-2}; units: N m^{-1}), is defined as the force in newtons acting along a line 1 m in length in the surface, and at right angles to the line. A molecule in the body of a liquid is acted upon by molecular attractions distributed symmetrically round the molecule. At an air–liquid interface, a molecule is only partially surrounded by other molecules and, as a result, experiences a net attraction into the bulk of the liquid. The effect of the attractive forces, known as the surface tension, is to tend to draw the molecules inward and make the liquid behave as if it were surrounded by an invisible membrane.

Additional surface is formed by moving molecules from the bulk of the liquid into the surface, doing work against the cohesive forces of the liquid. Thus, for an increase in area of $\Delta\sigma$ (at constant temperature and pressure) the increase in the surface free energy, ΔG_γ (units: J m^{-2}), is given by

$$\Delta G_\gamma = \gamma \Delta\sigma \quad \text{or} \quad \gamma = \left(\frac{\partial G}{\partial \sigma}\right)_{T,P}$$

Hence, γ may be considered as the work (J) required to produce an increase of area of 1 m^2; thus ΔG_γ and γ have the same numerical value but not the same dimensions or units.

As T increases, γ decreases, becoming zero at the critical temperature; many empirical equations have been put forward to account for this variation, e.g. that of McLeod:

$$\gamma = C(\rho_l - \rho_g)^4$$

where C is a constant for the particular liquid; and that of Sugden:

$$\gamma = \gamma_0(1 - T/T_c)^{1,2}$$

where γ_0 is a constant; this equation holds well for normal liquids over a wide range of temperatures.

For binary solutions, γ may increase slightly or decrease markedly with increase of concentration, owing to the adsorption of the component with the lower γ at the air–liquid interface (*Gibbs adsorption isotherm*, q.v.).

Methods of measurement of surface tension

In all experimental methods, great care must be taken to prevent the surface under study from becoming contaminated with grease, dirt etc.

(1) Capillary rise method, in which the height h which a liquid rises up a capillary tube of radius r is measured:

$$\gamma = \tfrac{1}{2}grh\rho$$

in which it is assumed that the liquid wets the glass. For very accurate work, the weight of the liquid in the meniscus (assumed to be spherical) is taken into account:

$$\gamma = \tfrac{1}{2}gh\rho(h + \tfrac{1}{3}r)$$

This is the only absolute method of measuring γ.

(2) The drop-weight method is based on the weight of liquid drops falling slowly from the tip of a thick-walled vertical tube. Experiment shows that all the suspended drop is not included in the falling drop and a correction factor, F, must be introduced:

$$\gamma = Fmg/r$$

where m is the average weight of each drop which falls from a tube of radius r. The correction factor, F, may be obtained from tables.

(3) The ring method depends on the fact that, when a horizontal ring is raised from the surface of a liquid, a portion of the liquid is raised with it. The maximum weight, m, of liquid supported by the ring is determined with a torsion balance. Again a correction factor, F', is included, since the cylinder of liquid held by the ring does not have vertical walls:

$$\gamma = \frac{mgF'}{4\pi R}$$

where R is the average radius of the ring.

This method allows rapid measurement of γ of small samples, although temperature control is difficult.

(4) The bubble pressure method depends on the fact that the pressure in a bubble the instant before it bursts is given by

$$P = g\rho h + 2\gamma/r$$

where h is the depth of the tube below the surface and r its radius. Experimental difficulties in this method have been overcome in the differential method, where the difference in pressure, Δp, required to blow bubbles at the ends of tubes of different radii is measured. Empirically, γ is given (to an accuracy of 0.1%) by

$$\gamma = A\,\Delta p\left(1 + 0.69\, r_{\mathrm{L}} \frac{g\rho}{\Delta p}\right)$$

where r_{L}/cm is the radius of the larger tube and A the apparatus constant, which must be determined by calibration.

See also J & P, P (Vol.II).

Surroundings
See System.

Symmetry number
The symmetry number, σ, is the number of indistinguishable positions into which a molecule can be turned by simple rigid rotation. In the homonuclear molecule A–A, in which the two ends of the molecule are identical, the position, after rotation through an angle of 180°, is identical with that held initially. In contrast, when a heteronuclear molecule A–B is similarly rotated, the same two positions are distinguishable. Thus each pair of positions, separated by 180°, represents two distinguishable positions for

A–B but not for A–A. The symmetry number or factor is introduced which makes the value of the *partition function* (q.v.) for rotation, Q_r, for a homonuclear molecule half the value of Q_r for an otherwise identical heteronuclear molecule.

For diatomic molecules: e.g. NO and CO, $\sigma = 1$; H_2, N_2, O_2, etc., $\sigma = 2$.

For polyatomic molecules:

(1) Linear molecules, treated as diatomic, e.g. HCN, N_2O, $\sigma = 1$; C_2H_2, CO_2, $\sigma = 2$.

(2) Spherical rotators, e.g. CH_4, CCl_4 with three identical moments of inertia, $I_A = I_B = I_C$, $\sigma = 12$.

(3) Spherical tops, e.g. NH_3, $CHCl_3$, where $I_A = I_B \neq I_C$, $\sigma = 3$.

(4) Asymmetrical tops, where I_A, I_B and I_C are all different (as an approximation $\bar{I} = \{I_A I_B I_C\}^{1/2}$), e.g. H_2O, $\sigma = 2$; C_6H_6, $\sigma = 12$.

System

The physical universe is, for convenience, divided into the system, i.e. the part under study, and the surroundings. A system may be simple or complex, homogeneous or heterogeneous. Boundaries may be real (e.g. walls of box) or imaginary.

An adiabatic system is one in which heat cannot pass across the boundaries, either inwards or outwards. Heat insulation does not exclude other interactions between the system and the surroundings.

A closed system is one in which matter cannot pass across the boundaries. Liquid and vapour sealed in a vessel constitute a closed system.

An open system is one in which matter can pass across the boundaries. Liquid in a vessel, bounded by the surface and the walls of the vessel, constitutes an open system, since liquid may escape as vapour, or a second substance can be added from the surroundings.

An isolated system is one in which there is no interaction between the system and its surroundings.

T

Temperature dependence of thermodynamic functions

See Enthalpy; Entropy; Free energy; Internal energy.

Ternary system

See Three-component system.

Thermal analysis

Thermal analysis is the study of cooling curves of simple substances or mixtures. When a substance, or mixture, in the liquid form (melt) is cooled at a uniform rate, the cooling curve (temperature–time) decreases regularly. If, however, there is a phase change at a given temperature, there will be a

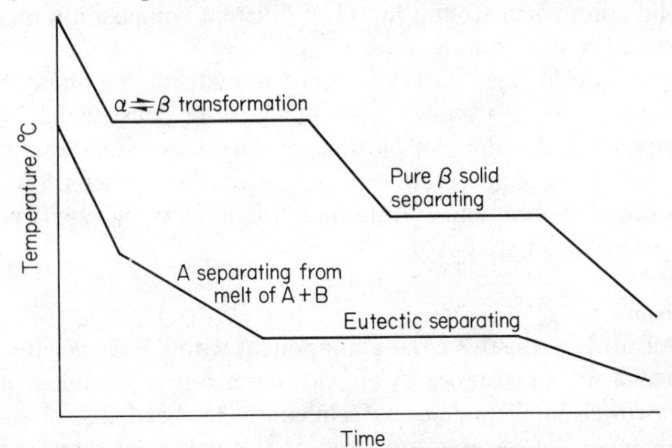

Figure T.1 Typical cooling curves.

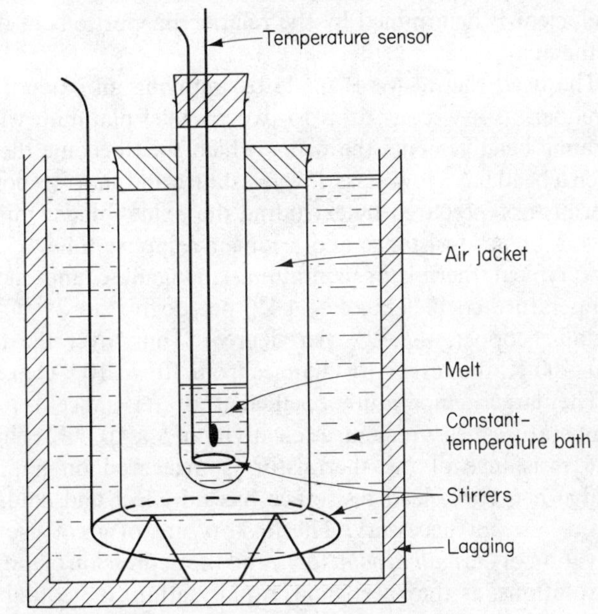

Figure T.2 Apparatus for determining cooling curves.

change in the rate of cooling, owing to heat evolved in the process, which produces a break in the cooling curve (figure T.1). When a solid compound, or a *eutectic mixture* (q.v.), separates from liquid of the same composition, heat is evolved and the temperature of the whole mass remains constant until it solidifies; allotropic transformations give a similar change. Separation of a solid component from a liquid of different composition merely results in a reduced rate of cooling.

From a series of such curves for known mixtures, the phase diagram for the *two-component condensed system* (q.v.) can be constructed.

The apparatus required for plotting cooling curves is shown in figure T.2. The nature of the temperature sensing device (thermometer, thermocouple, etc.) and the constant temperature bath (oil, ice, etc.) depends on the nature of the system under study.

Thermistor

A thermistor is a resistive circuit component with a large negative temperature coefficient of resistance. A typical thermistor is a stable, compact and rugged two-terminal ceramic-like semiconductor bead, disc or rod. Thermistors are made of various mixtures of the oxides of manganese, nickel, cobalt, copper, uranium, iron, zinc, titanium and magnesium; the temperature coefficient is determined by the relative proportions of the different oxides in the mixture.

The bead thermistor is made by applying an oxide mixture, containing a binder, as a viscous drop to two parallel platinum wires. On firing, the ceramic bead cements the wires, which then become the connecting leads. Such a bead has a low mass, low resistance and short response time. Rod-type thermistors, prepared by extruding the oxide–binder mixture before firing, have a higher resistance and a longer response time.

A typical thermistor containing manganese and nickel oxides has a temperature coefficient of -4.4% per degree at 25 °C (compare that of metallic copper, $+0.39\%$ per degree). Thus, over the temperature range 170–700 K, the resistivity changes from 10^5 to $10^{-2}\,\Omega$ m.

The large temperature coefficient of resistance is ideal for accurate temperature measurement, a sensitivity of 5×10^{-4} K being readily attained. The resistance of the thermistor is measured on a precision resistance bridge net-work; lead resistance has no effect and cold-junction compensators are not necessary. Thermistors are of great use in measuring the elevation of the *boiling point* (q.v.) and the depression of the *freezing point* (q.v.) of solutions, as they are small, rapidly attain the equilibrium temperature of the system and have a short response time.

Thermochemistry

Thermochemistry is that part of thermodynamics which deals with the heat changes accompanying a chemical reaction. Since the enthalpy change of a reaction depends on the temperature and pressure and on the states of the reactants and products, all of these must be specified in the normal balanced equation. Thus the heat evolved during the reaction between hydrogen and oxygen depends on the state of the water:

$$2H_2(g) + O_2(g) \longrightarrow 2H_2O(l) \qquad \Delta H^\ominus(298\ K) = -571.6\ kJ$$
$$2H_2(g) + O_2(g) \longrightarrow 2H_2O(g) \qquad \Delta H^\ominus(298\ K) = -483.6\ kJ$$

The difference between the amounts of heat liberated in the two reactions is due to the heat of vaporisation of water at 298 K, which may be calculated by subtraction of the two equations:

$$2H_2O(l) \longrightarrow 2H_2O(g) \qquad \Delta H^\ominus(298\ K) = 88.0\ kJ$$

i.e. the heat of vaporisation of water is 44.0 kJ mol^{-1} at 298 K. The designation (aq) indicates that such a state of dilution of the solution has been attained that no further enthalpy change occurs on dilution:

$$2KI(aq) + Cl_2(g) \longrightarrow 2KCl(aq) + I_2(s) \qquad \Delta H^\ominus(298\ K) = 224.0\ kJ$$

While the chemical equation must be balanced, it is often more convenient to include fractions of a mole of a substance, rather than multiplying out; this applies particularly to equations depicting the formation of 1 mole of a substance from its individual elements (see *enthalpy*).

See also Hess's law of constant heat summation.

Thermocouple

A thermocouple is a device which uses the voltage developed by the junction of two dissimilar metals to measure a temperature difference. Thermocouples are prepared by welding two wires of dissimilar metals together at one end. The junction is mounted in a sheath and the wires carefully insulated so that the only contact is at the junction. For most accurate work, two such metallic junctions are required, one, the cold junction, being kept at a standard temperature, usually 0 °C (in melting ice), and the other at the temperature to be measured (figure T.3). The thermoelectric e.m.f., measured directly with a potentiometer or digital voltmeter, is proportional to the difference in temperature between the two junctions. Reference books contain e.m.f.–temperature tables for a large number of metals which form suitable thermocouples. Any thermocouple prepared in the laboratory should be calibrated over the required range, in a well-stirred bath, the temperature

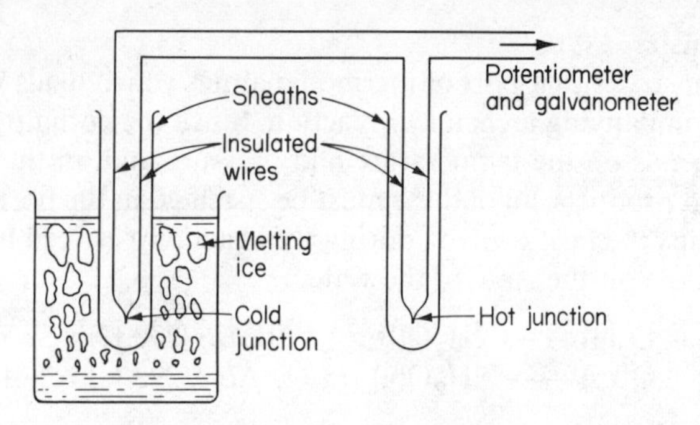

Figure T.3 Typical thermocouple with cold and hot junctions.

of which is measured with a standard mercury-in-glass thermometer or other standard thermocouple.

Since the thermoelectric e.m.f. does not depend on the size of the wires, small fine-wire thermocouples can be made which are ideally suited for the measurement of temperatures in small spaces where rapid changes of temperature occur.

In some commercial instruments using thermocouples as the sensing device, a cold reference junction is not required. Instead, a calibrated electrical compensator is provided so that the instrument reads temperature directly.

A thermopile, which consists of a number of thermocouples connected in series, provides a higher sensitivity, i.e. a greater temperature coefficient of e.m.f., than one couple alone.

Thermodynamic equations of state

$$\left(\frac{\partial U}{\partial V}\right)_T = T\left(\frac{\partial P}{\partial T}\right)_V - P \qquad \text{for } U = f(T, V)$$

$$\left(\frac{\partial H}{\partial P}\right)_T = V - T\left(\frac{\partial V}{\partial T}\right)_P \qquad \text{for } H = f(T, P)$$

These equations apply equally well to gases, liquids and solids, and are useful in interpreting equations of state for gases. For an ideal gas,

$$(\partial P/\partial T)_V = P/T; \text{ hence, } (\partial U/\partial V)_T = 0$$
$$(\partial V/\partial T)_P = V/T; \text{ hence, } (\partial H/\partial P)_T = 0$$

It is possible to evaluate $(\partial H/\partial P)_T$, which is related to the *Joule–Thomson coefficient* (q.v.) and $(\partial U/\partial V)_T$ from these equations for non-ideal gases.

Thermodynamic identities

From the *fundamental thermodynamic equations* (q.v.) the following are the more important identities which can be obtained:

$$T = \left(\frac{\partial U}{\partial S}\right)_{V,n_j} = \left(\frac{\partial H}{\partial S}\right)_{P,n_j}$$

$$P = -\left(\frac{\partial U}{\partial V}\right)_{S,n_j} = -\left(\frac{\partial A}{\partial V}\right)_{T,n_j}$$

$$S = -\left(\frac{\partial G}{\partial T}\right)_{P,n_j} = -\left(\frac{\partial A}{\partial T}\right)_{V,n_j}$$

$$V = \left(\frac{\partial G}{\partial P}\right)_{T,n_j} = \left(\frac{\partial H}{\partial P}\right)_{S,n_j}$$

Thermodynamic temperature scale

The thermodynamic temperature scale is the temperature scale which can be set up, by the *second law of thermodynamics* (q.v.), with reference to the properties of an ideal *heat engine* (q.v.), rather than to the properties of an ideal gas.

The ideal gas scale defines temperature in terms of the volume of an ideal gas at constant pressure, i.e. $T_2/T_1 = (V_2/V_1)_P$. The size of the unit on this scale is, by convention, the difference between the boiling and freezing points of water at atmospheric pressure, i.e. $T_{b,p,} - T_{f,p,} = 100$. On the thermodynamic scale the ratio of the two temperatures, θ_1 and θ_2, is given by the ratio of the heat absorbed at the two temperatures in a reversible *Carnot cycle* (q.v.), i.e. $\theta_2/\theta_1 = q_2/q_1$, and the size of the unit is, by convention, $\theta_{\text{b.p.}} - \theta_{\text{f.p.}} = 100$.

When an ideal gas is used as the working substance in a Carnot cycle, it can be shown that $T_2/T_1 = q_2/q_1$. It is thus apparent that $\theta_2/\theta_1 = T_2/T_1$, i.e. $\theta \propto T$; since, by definition, both scales coincide at two points, the temperature on the thermodynamic scale must be identical at all temperatures with that defined by a thermometer containing an ideal gas.

It is not necessary to choose two fixed points for the thermodynamic scale; it is preferable to choose a numerical value for one fixed point, the *triple point* (q.v.) of water ($T_{\text{t.p.}} = 273.1600$ K), and to define any temperature, T, by the relationship

$$T = 273.1600 \lim_{P \to 0} \{(PV)_T/(PV)_{T_{\text{t.p.}}}\}$$

See also K.

Thermogram
See Thermometric titration.

Thermometric titration
Thermometric titration is a titration in which the temperature change occurring during the titration is measured. In the simplest method, the titrant is added discontinuously from a burette into a Dewar flask containing the solution to be titrated, a stirrer and a *Beckmann thermometer* (q.v.). After each addition, the temperature rise is measured and the average temperature change per unit volume added is plotted against the total volume of the titrant added. A sharp change in this average value occurs at the equivalence point. The results are plotted either as the derivative curve just described or as a thermogram in which the total temperature rise is plotted against the volume of titrant added (figure T.4); the end-point being marked by a change in slope. This simple type of procedure, with an accuracy of about 1%, is suited to reactions which are complete and strongly exothermic ($\Delta H^{\ominus} \approx -120$ kJ mol^{-1}).

In modern titration calorimeters, the titrant is added continuously from a motor-driven syringe burette. The thermometer is replaced by a *thermistor* (q.v.) and the unbalanced potential from the bridge circuit, of which the thermistor is part, is continuously recorded. Thermometric titrations can now be achieved as rapidly as by any other instrumental method.

This improved recording method has changed classical reaction calorimetry. Instead of measuring the temperature change in a solution of one reactant by breaking within it an ampoule of a second reactant, it is now

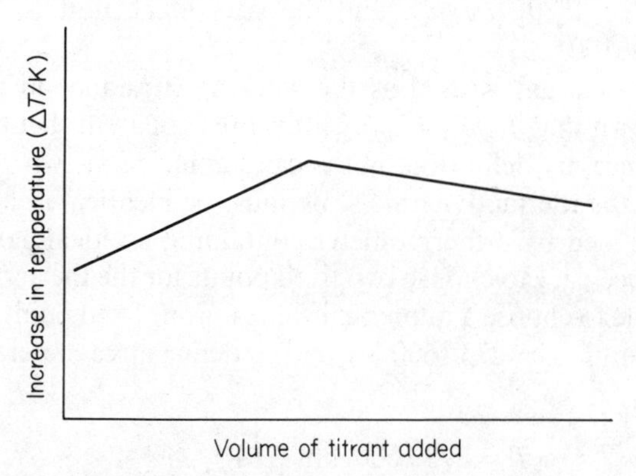

Figure T.4 Typical thermogram, showing increase of temperature as a function of the volume of titrant added.

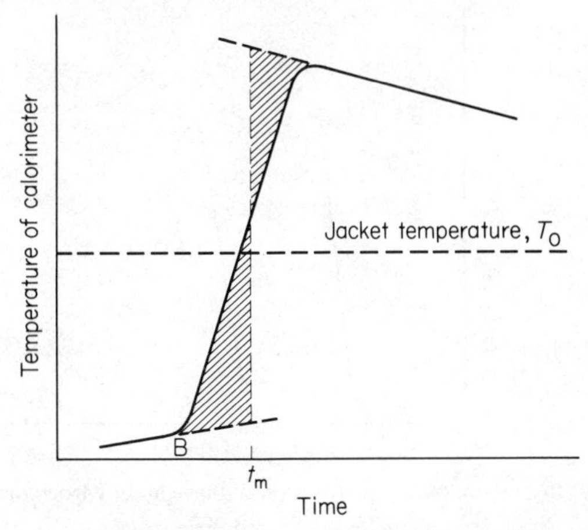

Figure T.5 Temperature–time curve for one addition of titrant (titrant added at B).

possible to add successive amounts of the second reactant at a carefully controlled temperature. The temperature rise can be found from an analysis of curves of the type shown in figure T.5. After electrical calibration, the heat change from each addition can be calculated. By successive additions up to and beyond the end-point, accurate enthalpy changes, referring to all stages of the reaction, can be calculated.

In the modern commercial instruments, great care has been taken to ensure good heat insulation and a small equilibration time, and in the design of heat exchangers through which the titrant must pass to attain a known temperature.

Thermometric techniques have been applied extensively to such problems as the following.

(1) Neutralisation reactions. Whereas the potentiometric or *electrometric titration*† of a weak acid with a strong base gives no discernible end-point, the thermometric titration gives as sharp an end-point as for the titration of a strong acid with a strong base (figure T.6). This is due to the fact that although the acid strengths differ greatly, the heats of neutralisation differ by only 30%.

(2) Estimation of sulphate, chloride, etc., by precipitation titrations.

(3) Compleximetric titrations, including EDTA titrations of solutions containing mixtures of metal ions.

(4) Titrations in non-aqueous solution; reactions between Lewis acids and Lewis bases.

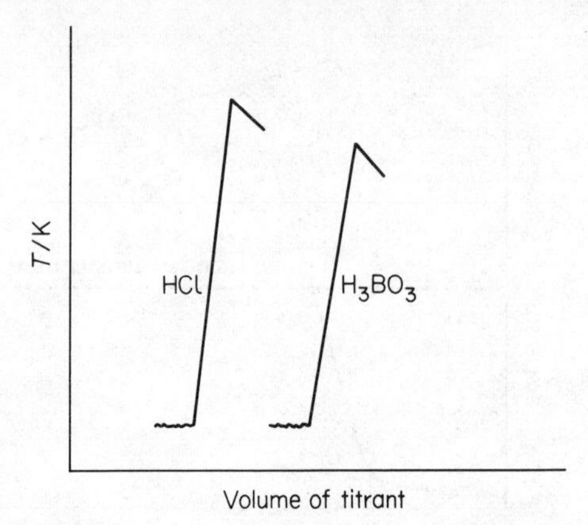

Figure T.6 Typical thermograms for the titration of boric and hydrochloric acids with sodium hydroxide solution.

(5) Study of proton equilibria in biologically important systems.
(6) Measurement of heats of dilution of acids and bases.
(7) Analysis of metal alkyls.

See also Thermometric Titrimetry, by H.J.V. Tyrrell and A.E. Beezer, Chapman and Hall (1968).

Third law of thermodynamics

Planck extended the *Nernst heat theorem* (q.v.) by making the additional postulate that the absolute *entropy* (q.v.) of a pure solid or a pure liquid approaches zero at 0 K, i.e. $\lim_{T \to 0} S = 0$. Since solutions and supercooled liquids (even of a pure substance) retain positive entropies as the temperature approaches 0 K, Lewis and Randall proposed the following statement of the third law: 'If the entropy of each element in some crystalline state be taken as zero at the absolute zero of temperature, every substance has a finite positive entropy; but at the absolute zero of temperature the entropy may become zero, and does so become in the case of perfectly crystalline substances.'

The entropy is related to the thermodynamic probability by

$$S = k \ln W$$

thus the more randomly (orderly) the molecules are arranged, the greater (smaller) are W and S. If the system is completely ordered, $W = 1$ and $S = 0$. This is true of a perfectly crystalline solid as T approaches 0 K and the lattice sites are completely fixed; on this basis, a glass or supercooled liquid would be expected to have a residual entropy at 0 K.

As with the other laws, the third law is an expression of our experiences with nature, and it provides a method for obtaining the thermal value of the entropy of a system:

$$S^{\ominus}(T) = S_0^{\ominus} + \int_0^T C_p^{\ominus} \frac{\mathrm{d}T}{T} + \sum \frac{\Delta H(\text{trans})}{T(\text{trans})}$$

When the substance under consideration fulfils the necessary criterion of the third law, $S_0^{\ominus} = 0$, the integral may be evaluated to obtain the absolute thermal entropy of the system. Knowing $S^{\ominus}(T)$ for each of the reactants and products in a given reaction and also $\Delta H^{\ominus}(T)$ for the reaction, the value of $\Delta G^{\ominus}$ may be calculated.

All attempts to attain lower and lower temperatures lead to the general statement that the absolute zero of temperature is unattainable.

An elegant test of the third law involves the use of two crystalline forms of a single substance. If a substance does exist in two perfectly crystalline forms, then the entropy change at 0 K for the transition should be zero. $\Delta S_0^{\ominus}$ for the transition from rhombic to monoclinic sulphur may be calculated from heat capacity data and the heat of transition at 368 K:

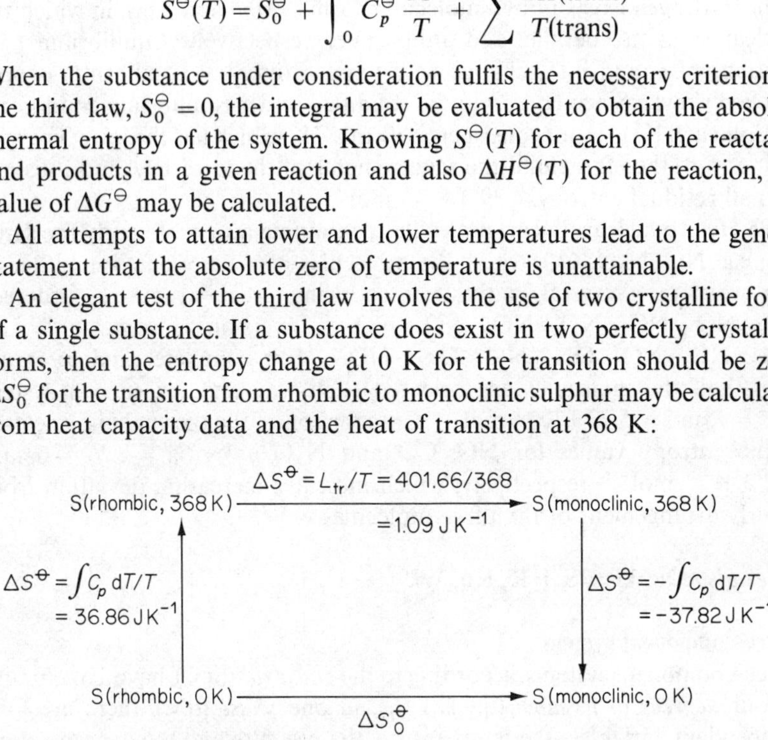

S(rhombic, 368 K) → S(monoclinic, 368 K): $\Delta S^{\ominus} = L_{tr}/T = 401.66/368 = 1.09\ \mathrm{J\,K^{-1}}$

S(rhombic, 0 K) → S(rhombic, 368 K): $\Delta S^{\ominus} = \int C_p\, \mathrm{d}T/T = 36.86\ \mathrm{J\,K^{-1}}$

S(monoclinic, 368 K) → S(monoclinic, 0 K): $\Delta S^{\ominus} = -\int C_p\, \mathrm{d}T/T = -37.82\ \mathrm{J\,K^{-1}}$

S(rhombic, 0 K) → S(monoclinic, 0 K): $\Delta S_0^{\ominus}$

whence

$$\Delta S_0^{\ominus} = 36.86 + 1.09 - 37.82 = -0.13\ \mathrm{J\,K^{-1}\,mol^{-1}}$$

Within the limits of accuracy of the thermal data, $\Delta S_0^{\ominus} = 0$, as predicted by the third law.

Solutions and glasses are excluded from the third law. Thus, for the transition,

$$\text{glycerol (cryst, 0 K)} \longrightarrow \text{glycerol (glass, 0 K)}$$

$\Delta S_0^{\ominus} = 19.3\ \mathrm{J\,K^{-1}\,mol^{-1}}$, calculated from heat capacity data.

Some thermal entropy values for gases are significantly less than the *spectroscopic entropy* (q.v.) values. In all such cases, a random configuration can exist in the solid at 0 K, producing a residual entropy. Such substances are of two types.

(1) Homonuclear diatomic molecules, whose nuclei possess a resultant spin. Hydrogen exists in two molecular forms, *ortho* and *para*, in which the nuclear spins are parallel and anitparallel, respectively. Equilibrium proportions of *o*- and *p*-H_2 change with temperature, but equilibrium is only reached slowly. Since H_2 cannot be solidified under equilibrium conditions, the solid at 0 K does not consist entirely of *p*-H_2 (which has the lower energy). It does not, therefore, fulfil the requirements of the third law, and thus has a small residual entropy, 4.39 J K^{-1} mol^{-1}.

(2) Heteronuclear (linear) molecules containing atoms of nearly the same size, e.g. NO, N_2O, CO, where the molecules may be oriented in either of two ways in a crystal, without introducing great stress, e.g. ordered: NO—NO—NO—NO—NO—NO—NO—; inversion of some molecules gives: NO—NO—ON—NO—ON—NO—. For complete random structures, half the molecules are oriented each way and for such a system $\Delta S^{\ominus} = R \ln 2 = 5.73$ J K^{-1} mol^{-1}. Comparison of the thermal and spectroscopic entropy values for NO, CO and N_2O gives $S_0^{\ominus} = 2.76$, 4.6 and 4.77 J K^{-1} mol^{-1}, respectively, which indicates increasing deviation from orderly arrangement of the three molecules.

See also Dic, G & S, I, K, Kn, Wi.

Three-component system

Three-component systems, according to the *phase rule* (q.v.), have a maximum of four *degrees of freedom* (q.v.). Thus, in one *phase* (q.v.), there are four independent variables: the temperature, the pressure and the concentration of two *components* (q.v.). Since it is not possible to represent four variables graphically, some simplification is needed. Normally the pressure is fixed at atmospheric and one degree of freedom is surrendered; the remaining variables may then be plotted on a three-dimensional graph (figure T.7), the concentration variables on an equilateral triangle, forming the base, and the temperature on the vertical axis. Further simplification is achieved by plotting horizontal sections of this at fixed T. The equilateral triangle, on account of its geometry, is most convenient for the representation of concentrations. At any point P (figure T.8), if the lines Pa, Pb, and Pc are drawn parallel to AB, BC and CA, respectively, then Pa + Pb + Pc = AB;

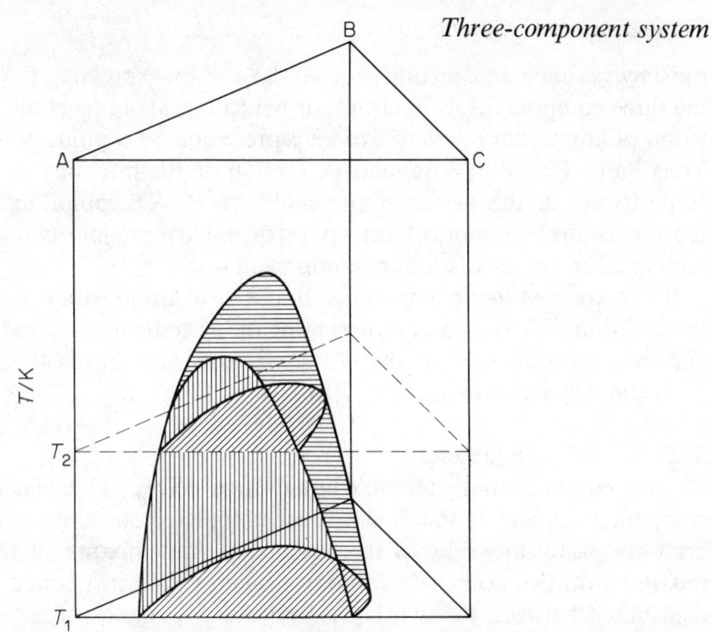

Figure T.7 Three-dimensional representation of the equilibrium curve for components A, B and C, showing sections at T_1 and T_2.

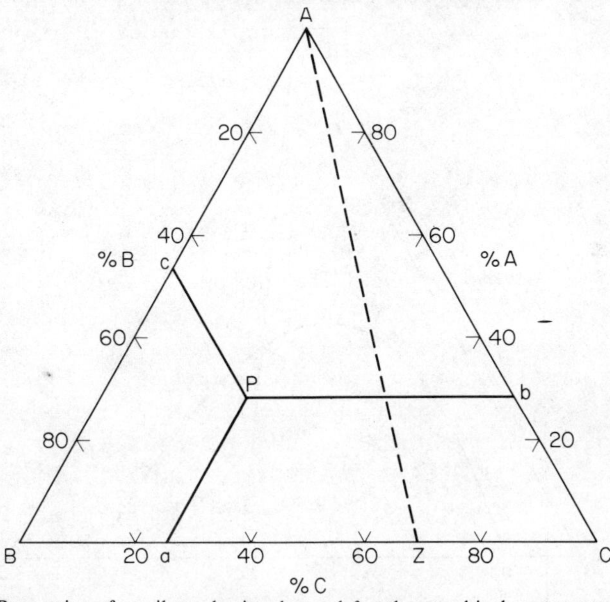

Figure T.8 Properties of equilateral triangle used for the graphical representation of three-component systems.

thus, by taking a side as unity (or 100%) and by expressing the amounts of the three components as fractions (or percentages) of the whole, the composition of any ternary system can be represented by a point in the diagram. Thus Pa (=Cb), Pb (=Ac) and Pc (=Ba) are the fractions of A, B and C, respectively, in the system represented by P. Any point in the triangle represents three components; any point on a side represents two components; and an apex represents a pure component.

If a mixture of two components, B and C, of composition Z is diluted by the addition of A, the point representing the system moves along the line ZA, and the ratio of B to C in any of the resulting mixtures is always the same as in the original mixture (ZC : ZB).

Systems of three liquids

(1) One pair of partially miscible liquids (figure T.9). At a given temperature, components A and B and A and C are completely miscible, while B and C are only partly miscible. In the absence of A, a mixture of B and C will separate into two conjugate solutions, b and c. When A is added, it will be distributed between the two layers and cause a change of compositions to b′ and c′. These layers in equilibrium, known as conjugate ternary solutions, are joined by a *tie line* (q.v.), b′ c′; this, sloping upwards to the right, indicates

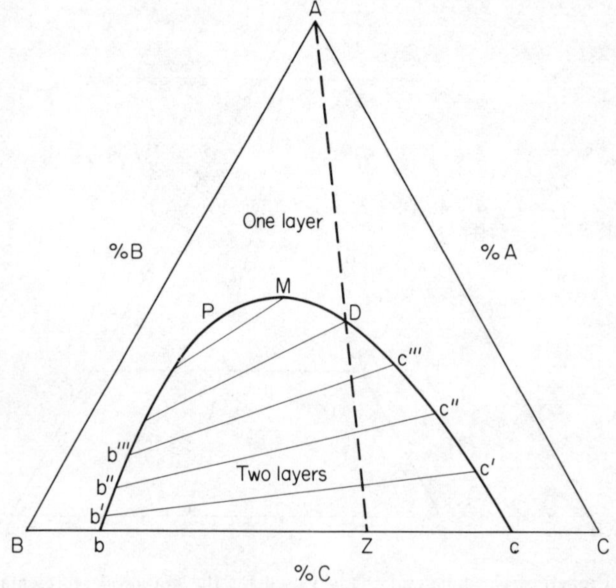

Figure T.9 System of three liquids, one pair being partially miscible.

that A is more soluble in the layer richer in C. A series of such experimental determinations of the compositions of conjugate solutions provides data for plotting the closed curve (b, b′, b″ ... P, M ... c″, c′c) known as the binodal curve. The compositions of the two layers approach each other as seen from the shortening of the tie lines, until at P, the plait point, they are equal and only one layer exists. The binodal curve has a maximum at M which does not, in general, coincide with P. Any point outside the curve represents one liquid layer, $f' = 2$ (apart from T and P), so the system is completely defined by fixing the composition of two components. Any mixture of composition lying within the curve separates into two conjugate solutions, $f' = 1$.

The addition of A to a binary system of composition Z causes the composition of the system to move along ZA. From the positions where this cuts the tie lines, it is obvious that the amount of left-hand layer (i.e. that richer in B) becomes less and less, until at D it disappears, and only one layer is left from D to A.

Examples: acetic acid, chloroform and water; acetone, water and phenol.

With increase in temperature, the mutual solubilities of the components

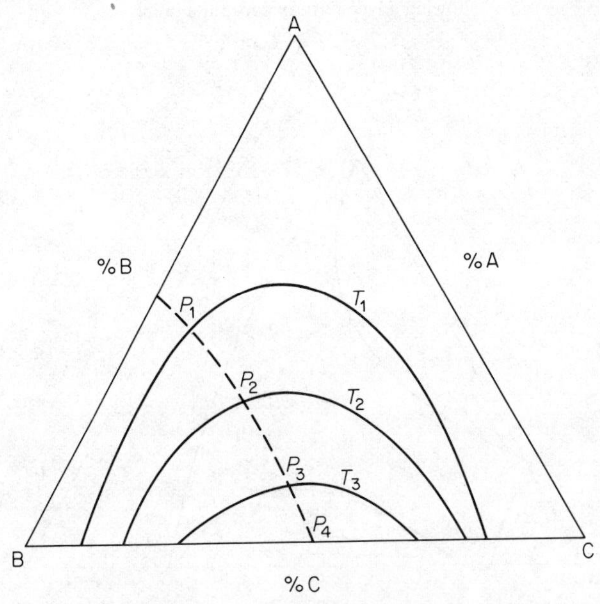

Figure T.10 Effect of temperature on a system of three liquids, one pair partially miscible, ($T_3 > T_2 > T_1$), showing a binary consolute point.

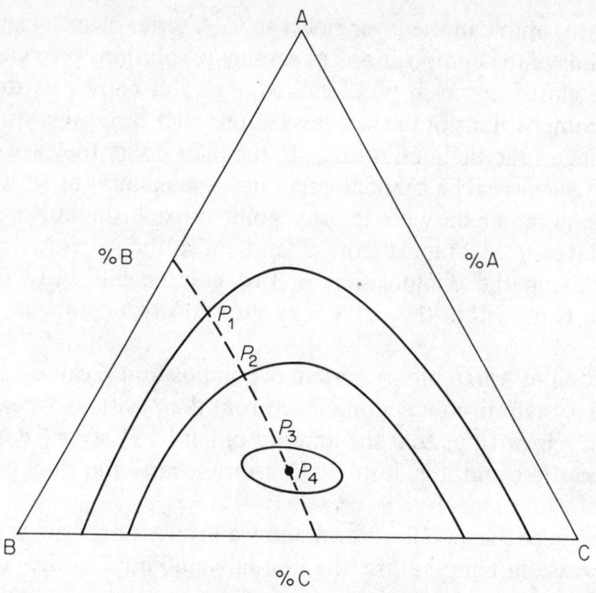

Figure T.11 Effect of temperature on a system of three liquids, one pair partially miscible, showing a true ternary consolute point.

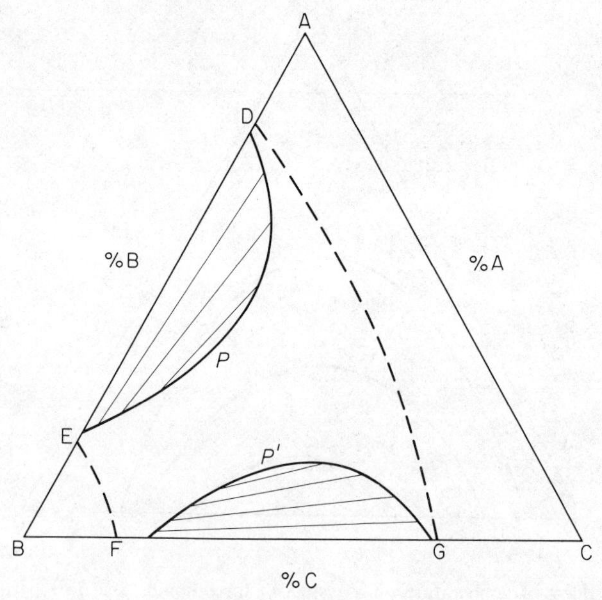

Figure T.12 System of three liquids with two pairs of partially miscible liquids.

usually increase (very few decrease) and, hence, the binodal curves progressively enclose smaller areas, which finally disappear to give a binary consolute point (figure T.10) or a true ternary consolute point (figure T.11). The final consolute point P_4 corresponds to the peak of the solid in figure T.7.

(2) Two pairs of partially miscible liquids (figure T.12). When A and B as well as B and C are partially miscible, there are two binodal curves each with its own tie lines and plait point. At lower temperatures, the two curves may coalesce to give a continuous band (DEFG).

Examples: water, ethanol and butanol; methanol, heptane and nitrobenzene.

(3) Three pairs of partially miscible liquids. Three separate binodal curves are possible; points within each represent a pair of conjugate ternary liquids in equilibrium (figure T.13). For less miscible liquids, or at lower temperatures, the curves merge to give figure T.14, in which it is possible to obtain areas of one, two or three layers as shown. The triangular area DEF represents three liquid layers in equilibrium, of compositions given by the points D, E and F.

Example: water, ether and succinic nitrile.

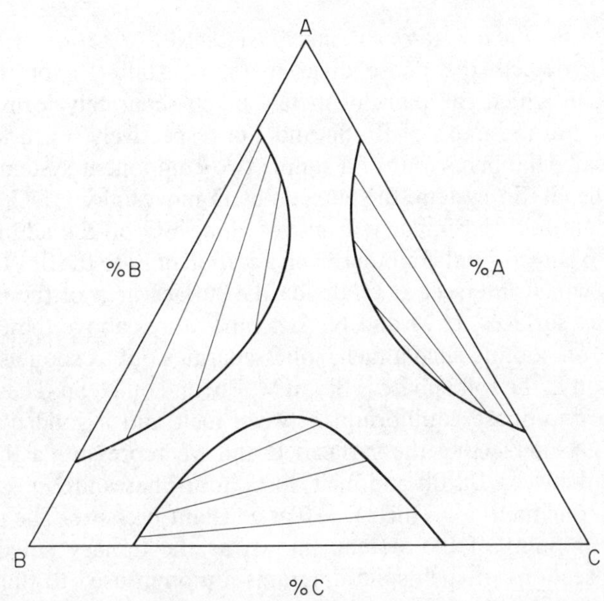

Figure T.13 System of three liquids; three pairs of partially miscible liquids.

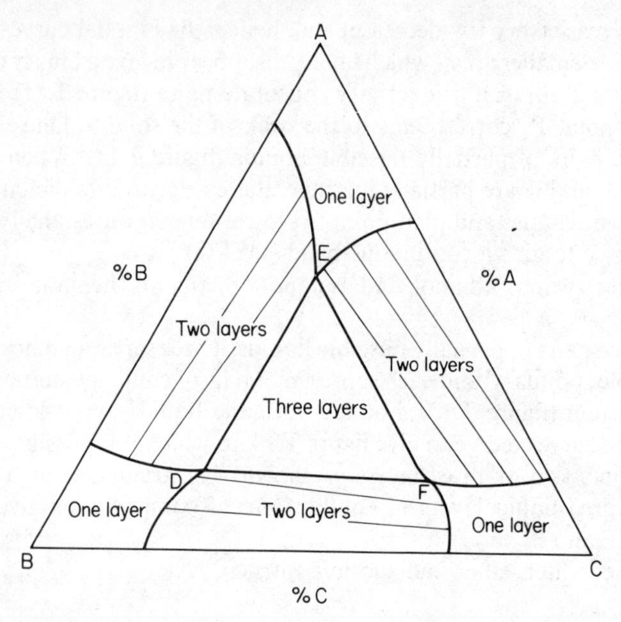

Figure T.14 System of three partially miscible pairs of liquids which can exist in three layers.

Systems with solid phases (ternary eutectic systems)
Figure T.15 depicts the phase diagram (at constant P) for the system Bi, Sn, Pb, in which the pairs of metals taken separately form eutectics. A, B and C are the m.p.s of Bi, Sn and Pb, respectively. Each face of the prism indicates the behaviour of a simple two-component system. As Pb is added to the Bi–Sn system, the eutectic at D moves along DG according to the amount added. Similarly, F moves along FG on the addition of Sn to the Bi–Pb eutectic and E along EG on addition of Bi to the Sn–Pb system; these lines, which intersect at G, divide the surface area of the prism into three distinct surfaces, L, M and N. At temperatures above these, only the melt exists; on cooling such a melt, solid separates out as soon as a surface is reached; on L, the solid phase is Bi; on M, Pb; and on N, Sn. These surfaces represent a two-phase equilibrium between melt and a solid component. The line FG, intersecting the surfaces L and M, represents a three-phase equilibrium between Bi, Pb and melt; at G, four-phases are in equilibrium (three solid, one melt) for which $f' = 0$ at constant pressure. The point G is the invariant point of the system, known as the ternary eutectic point. Isothermal sections of such solid diagrams are often used to illustrate the behaviour of these systems.

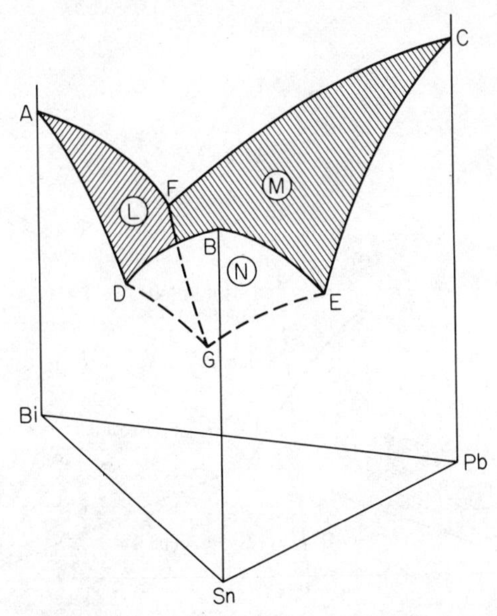

Figure T.15 The ternary eutectic system bismuth, tin and lead.

Solutions of two salts with a common ion
When two salts which have an ion in common are dissolved in water, a three-component system is formed. The phase diagrams for such systems are obtained from a study of the solubility of each salt in aqueous solution containing varying amounts of each salt at constant temperature (above 0 °C) and pressure. Various mixtures are prepared and thermostatted until equilibrium is attained; the wet solid is removed from the supernatant liquid and the composition of both solid and liquid phases determined. Schreinemaker's wet residue method is used to discover the nature of the solid phase in equilibrium with the various solutions. This is based on the fact that the *tie line* (q.v.), drawn through the composition of the wet solid r_1 and liquid l_1 in equilibrium, must pass through the composition of the solid phase (figure T.16), and, moreover, since several solutions may have the same solid phase, all the tie lines for such solutions (l_2r_2, l_3r_3, etc.) must pass through a common point, i.e. the composition of the solid phase.

In the simplest of all systems (figure T.16) in which components S_1 and S_2 separate out and in which there is no compound formation, the isothermal evaporation of an unsaturated solution l, must pass along the line Wl produced to t (W is water). When the solubility curve AC is reached at m,

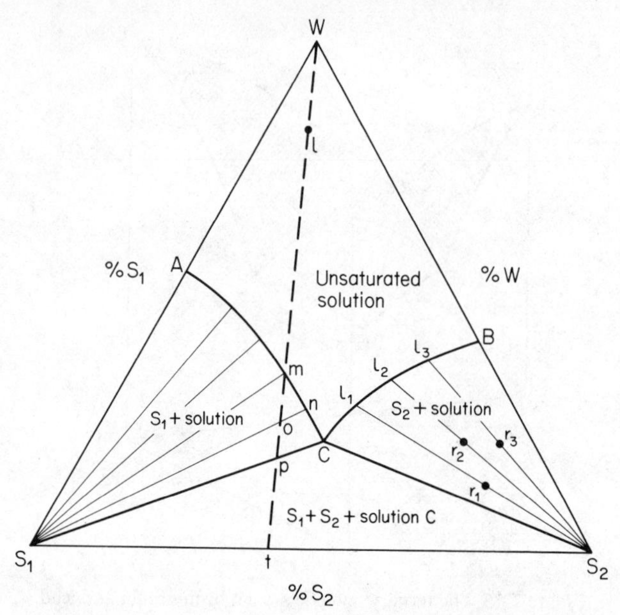

Figure T.16 Three-component system of two salts with a common ion and water.

solid S_1 begins to separate, and the liquid becomes richer in S_2 and passes along the line mC. When the composition of the whole system is at o, solid S_1 is in equilibrium with solution of composition n; at p, the composition of the solution has reached C, and solid S_2 now starts to deposit. C is an invariant point ($f' = 0$), and the composition of the liquid phase remains constant while S_1 and S_2 are separating out; when t is reached, the liquid phase has disappeared, and the system consists of the two solids in the proportions originally present in the unsaturated solution at l.

Example: water, NaCl and NH_4Cl.

When hydrate formation occurs (figure T.17), its composition lies along WS_1 and is given by $S_1 \cdot xH_2O$ (D). E is the solubility of the hydrate in water and EC gives the solubility of the hydrate in solutions containing S_2. At C, the isothermal invariant point, the solution is saturated with both $S_1 \cdot xH_2O$ and S_2. Example: water, Na_2SO_4 ($x = 10$), NaCl at 15 °C. Under some temperature conditions, the hydrate and anhydrous salt may appear (figure T.18); three solubility curves are obtained (one for each solid) and two invariant points, F and G. Example: water, Na_2SO_4 ($x = 10$) and NaCl at 25 °C. At temperatures above 32 °C, the decahydrate does not exist and figure T.18 simplifies to figure T.16.

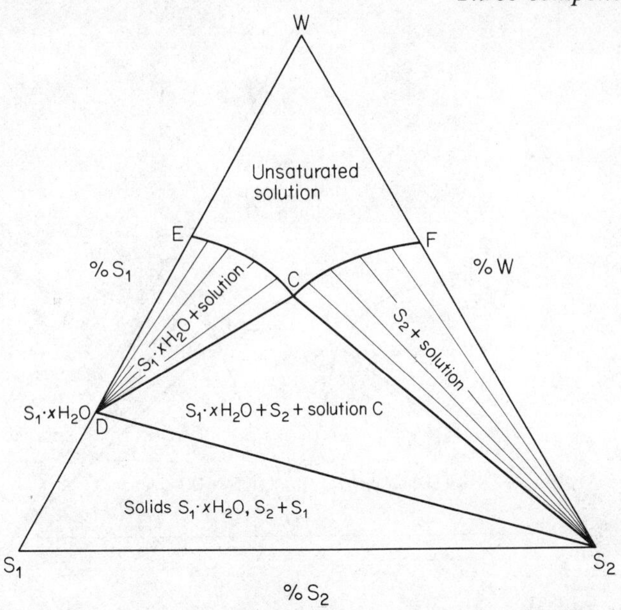

Figure T.17 Two salts with a common ion and water, showing the formation of a hydrate.

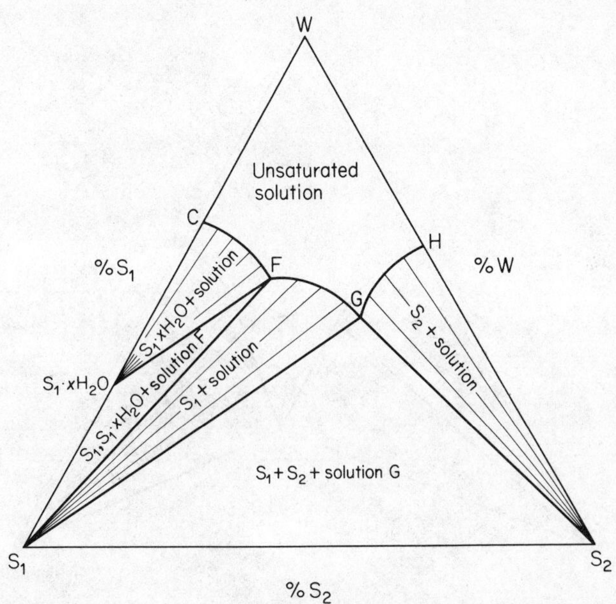

Figure T.18 Two salts with a common ion and water, showing the appearance of a hydrate and pure components.

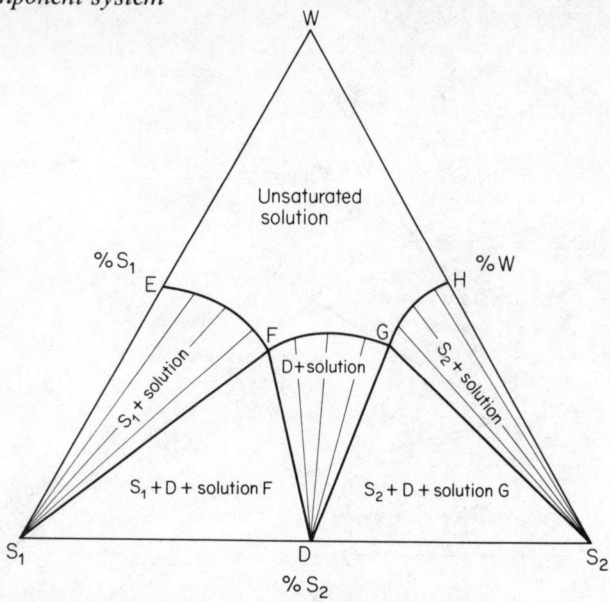

Figure T.19 Two salts with a common ion and water, showing the formation of a double salt.

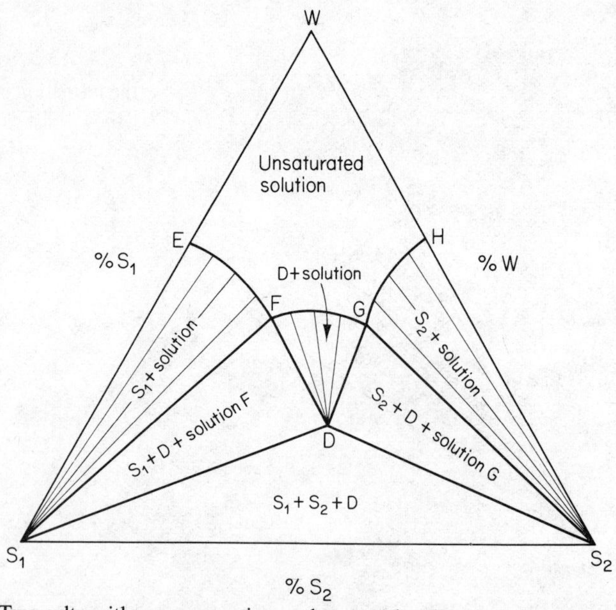

Figure T.20 Two salts with a common ion and water, showing the formation of a hydrated double salt.

The two salts may form a double salt or complex compound (figure T.19), e.g. water, NH_4NO_3 and $AgNO_3$, the point D corresponding to $NH_4NO_3 \cdot AgNO_3$. The double salt may be hydrated (figure T.20), e.g. water, $MgSO_4$ and Na_2SO_4, the point D corresponding to $MgSO_4 \cdot Na_2SO_4 \cdot 4H_2O$.

See also Bo, F & C, F & J, M & P.

Tie line

A tie line on a phase diagram (figure T.21) connects the composition of the two phases in equilibrium at a given temperature (pressure) in a phase

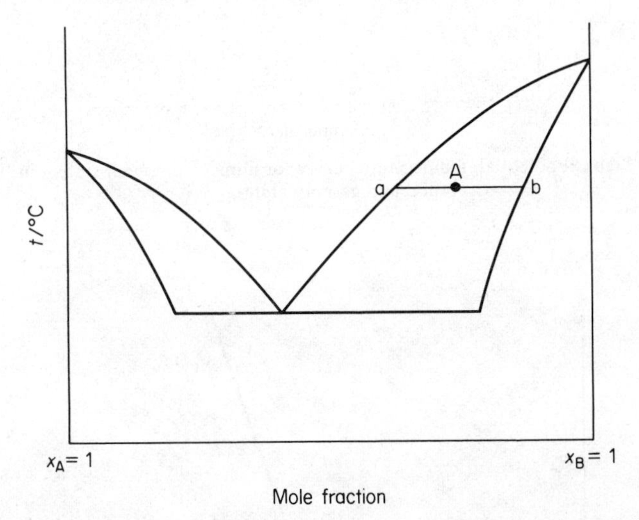

Figure T.21 Phase diagram for two-component system, illustrating a tie line.

diagram determined at constant pressure (temperature). A mixture of composition A separates into phases of composition a and b such that:

$$\frac{\text{Amount of phase a}}{\text{Amount of phase b}} = \frac{\text{Ab}}{\text{Aa}}$$

See also Binary liquid mixture; Three-component system; Two-component system.

Triple point

If phase equilibrium in a *one-component system* (q.v.) is extended to three phases, then the *chemical potential* (q.v.) of the component must be the same

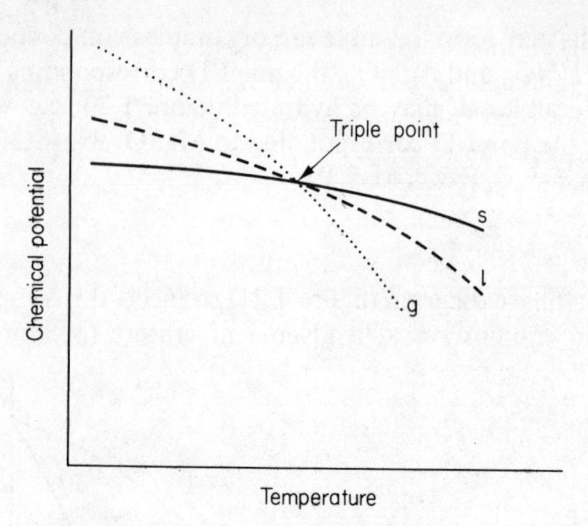

Figure T.22 Chemical potential–temperature curves for a one-component system in the solid, liquid and gaseous states.

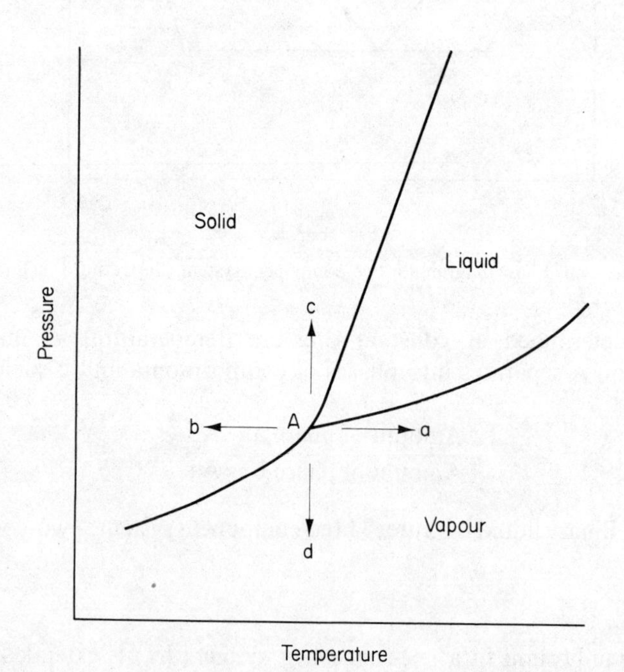

Figure T.23 Vapour pressure–temperature curves for a one-component system.

in all three phases. Intersection of the three chemical potential–temperature curves of the liquid, solid and vapour will occur at a unique temperature and pressure for each substance (figure T.22). The triple point is the point at which any three phases (e.g. $S_\alpha LV$, $S_\alpha S_\beta V$, $S_\alpha S_\beta L$, $S_\beta LV$) co-exist in equilibrium. For such a system in equilibrium, $c = 1$, $p = 3$ and, hence, from the *phase rule* (q.v.), $f = 0$. The point is thus invariant: T and P are fixed for the given system (figure T.23). So long as three phases are present, no change in T or P of the system can occur, but only changes in the relative amounts of the three phases. Continued addition of heat (line Aa) at constant P will ultimately result in the disappearance of the solid and liquid forms and the formation of a bivariant vapour system; similarly, continual removal of heat (Ab) results in the formation of the bivariant solid. Increase in pressure at constant temperature (Ac) results in the formation of either the solid or liquid phase, depending on the slope of the P–T curve (see *water system*), while reduction of pressure (Ad) results in the formation of the vapour phase. The total possible number of triple points in a one-component system, given by

$$\frac{p(p-1)(p-2)}{1 \times 2 \times 3}$$

increases rapidly as the number of possible phases increases. In the *sulphur system* (q.v.), $p = 4$, (i.e. S_α, S_β, liquid and vapour) and there are four triple points.

Trouton's rule

Trouton's rule states that the molar heat of vaporisation/J mol^{-1}, divided by the normal boiling point, is approximately 90 for most liquids. This is equivalent to the statement that the entropy of vaporisation is the same for most liquids:

$$\Delta S(\text{vaporisation}) = L_e/T \approx 90 \text{ J K}^{-1} \text{ mol}^{-1}$$

The law is valid for a large number of non-polar liquids with very different boiling points. Serious disagreement is shown by liquids which are associated by hydrogen bonding in the liquid state: e.g. water, 108.5; ethanol, 110. Hydrogen bonding causes considerable restriction on the freedom of movement of molecules in the liquid, and, hence, a lower entropy in the liquid than is found with non-associated liquids. Carboxylic acids—e.g. formic acid, 64.4, and acetic acid, 62.2—show the opposite deviation, again due to hydrogen bonding, but here the bonds result in the formation of dimers which persist to an appreciable extent in the vapour phase.

Trouton's rule

Assuming that, in the liquid phase, there is a volume in which the liquid molecules are free to move, a volume known as the free volume, $V(\text{free})$, then

$$\Delta S(\text{vaporisation}) = S(\text{g}) - S(\text{l}) = R \ln \frac{V(\text{vapour})}{V(\text{free})} \approx 90$$

gives $V(\text{vapour})/V(\text{free}) \approx 10\,000$; since 1 mole of vapour occupies about 0.2 to 0.3 m^3, $V(\text{free})$ must be about $2\text{–}3 \times 10^{-6}$ m^3. As a typical molar volume of a liquid is about 10^{-4} m^3, this means that 2–3% of the liquid volume is free volume in which the liquid molecules can move.

Two-component condensed system

In the consideration of solid–liquid equilibria, the effects of pressure change are usually negligible; when the vapour phase is ignored, the system is termed condensed and one *degree of freedom* (q.v.) is removed; the *phase rule* (q.v.) becomes

$$p + f' = c + 1$$

The remaining variables—temperature and concentration—are plotted in equilibrium diagrams in the usual way. The presence of only one phase, $f' = 2$, is represented by an area; the equilibrium between two phases, $f' = 1$, is represented by a line; and the equilibrium between three phases, $f' = 0$, is represented by an invariant point.

The more important types of equilibria between two components, A and B, which are completely miscible in the liquid state, are as follows.

(1) Simple eutectic systems in which only pure A and B crystallise from the solution or melt. The addition of a small amount of B lowers the melting point of A (figure T.24); AC can be considered either as the freezing point curve of the liquid or the solubility curve of A, and represents the temperature at which various liquid mixtures are in equilibrium with pure solid A. Similarly for BC. Where AC and BC meet, solid A, solid B and liquid are in equilibrium ($f' = 0$); this invariant point is the eutectic temperature. When a melt of composition p is cooled (figure T.24b), the temperature falls fairly rapidly at first along pq. At q, solid A is in equilibrium with liquid of composition q, and, on further cooling, solid A separates out (the rate of cooling is reduced since heat is evolved) and the composition of the liquid passes along qC, becoming richer in B. When the temperature reaches r, the liquid has the eutectic composition, and so solids A and B separate out (in a fixed ratio), the temperature remaining constant, rs, owing to the evolution of heat, until all the liquid has solidified. The length of the

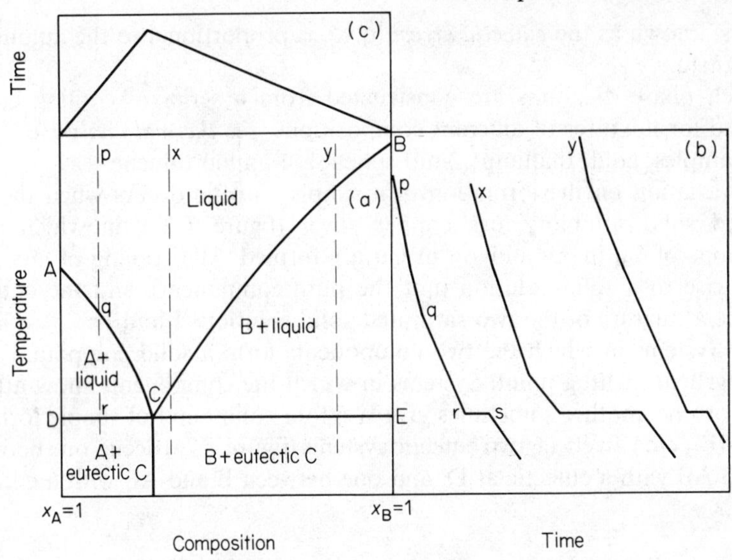

Figure T.24 Two-component condensed system; no compound formation. (a) Phase diagram; (b) cooling curves; (c) duration of eutectic arrest.

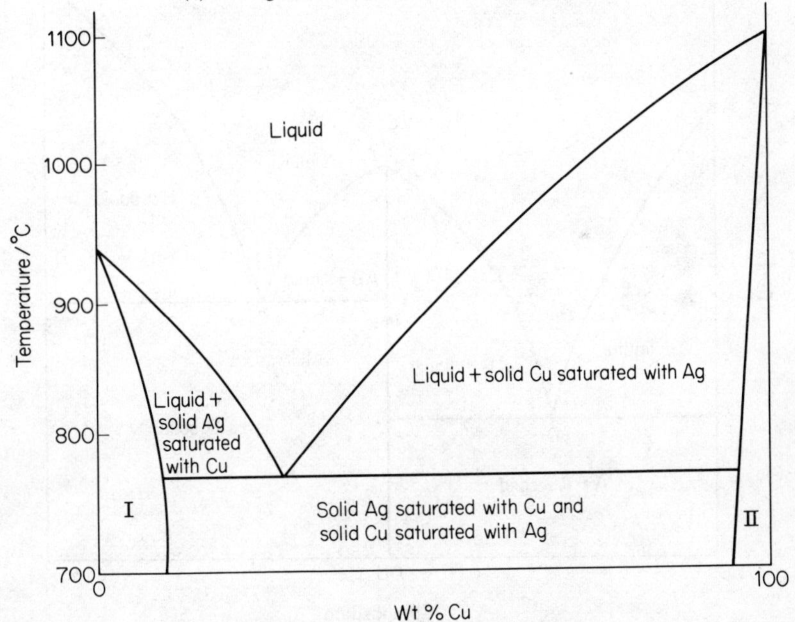

Figure T.25 Freezing point diagram for the silver–copper system. I, Solid Ag with dissolved Cu; II, solid Cu with dissolved Ag.

line rs, known as the *eutectic arrest* (q.v.), is proportional to the amount of *eutectic* (q.v.).

Such phase diagrams are constructed from a series of cooling curves plotted for mixtures of different compositions (see *thermal analysis*).

Examples: gold–thallium; 2-nitrophenol–4-amino toluene.

A variation on the formation of a simple eutectic occurs when there is limited solid solubility, e.g. copper–silver (figure T.25), in which solid solutions of Cu in Ag and Ag in Cu are formed. The cooling of any melt gives rise to a solid solution (not the pure component), and the eutectic will be a mixture of the two saturated solid solutions I and II.

(2) Systems in which the two components form a solid compound with a congruent melting point. Systems in which the components show attraction for one another sometimes give rise to a compound of simple formula (AB, AB_2, etc.). In effect, two eutectic systems (figure T.26) occur, one between A and AB with a eutectic at D, and one between B and AB with a eutectic

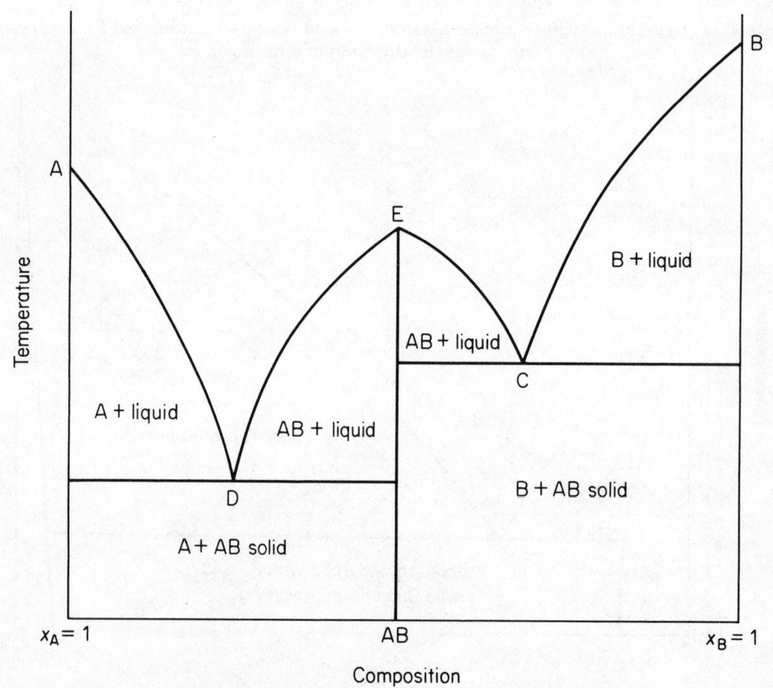

Figure T.26 Freezing point–composition diagram for a two-component system which forms a compound with a congruent melting point.

at C. The m.p., E, of AB may be above or below the m.p. of A or B; the composition of the solid, but not of the eutectic mixtures, is independent of the temperature and pressure. The stability of the compound may be inferred from the shape of the curve around E. If AB is stable, the curve will be sharp; but if unstable, it will be flatter, owing to the depression of the f.p. of AB as a result of the presence of components A and B from the dissociation of the compound.

Examples: Al–Mg (Al_3Mg_4); Au–Sn (AuSn); salt hydrates H_2SO_4–H_2O ($H_2SO_4 \cdot H_2O$, $H_2SO_4 \cdot 2H_2O$, $H_2SO_4 \cdot 4H_2O$); and iron (III) chloride–water (several hydrates).

(3) Systems in which the two components form a compound with an incongruent melting point. This complication occurs when the compound formed is not stable in the liquid state and decomposes below its m.p. The solid cannot exist in equilibrium with liquid of the same composition. The meritectic melting can be represented by an equation of the form

$$\text{Solid AB} \rightleftharpoons \text{Liquid} + \text{Solid B}$$

If the melt of composition a (figure T.27) is cooled, solid B starts to separate

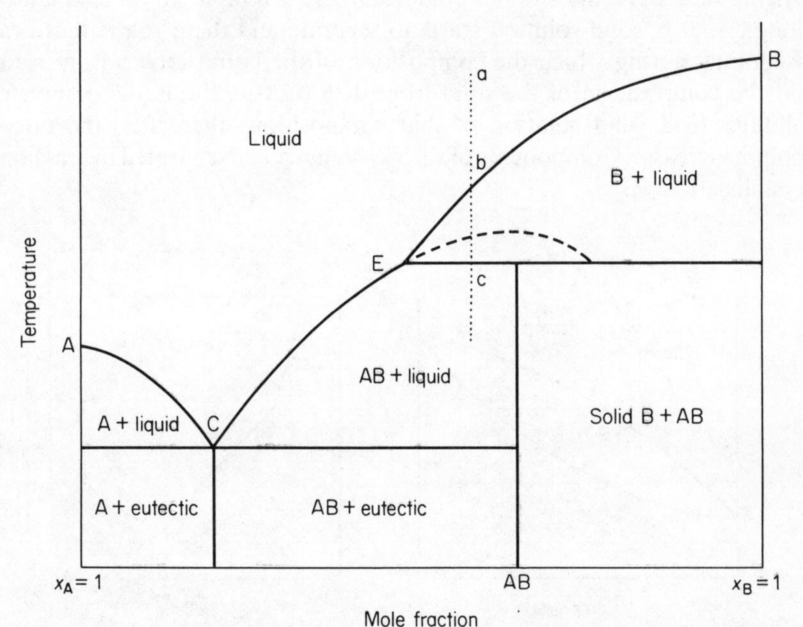

Figure T.27 Freezing point–composition diagram for a two-component system which forms a compound with an incongruent melting point.

at b, and the composition of the liquid falls along BE. At c, AB is stable and starts to separate ($p = 3$, $f' = 0$); hence, at this invariant point, the temperature remains constant while solid B disappears (i.e. right to left in the above equation). Further cooling leads to the separation of more AB and eventually the formation of eutectic mixture C of solids A and AB.

Examples: benzene–picric acid; tin–gold; salt hydrates such as $Na_2SO_4 \cdot H_2O$.

(4) Systems in which the two components are completely miscible in the solid state and form solid solutions (cf. *binary liquid mixture*). Complete miscibility of two solid phases results when the atoms of the two components are about the same size and can substitute for one another in the lattice. There are three types:

(a) Those in which there is a continuous series of solid solutions (figure T.28). The f.p.s. of all mixtures lie between those of the pure components; the lower curve is the m.p. of the solid solution and the upper curve is the composition of the liquid in equilibrium with the melting solid at the same temperature. Since at no time are there more than two phases, there can be no invariant point, and, hence, no horizontal portion on the cooling curve. When a melt of composition a is cooled, there is a rapid fall in temperature along ab; at b, solid solution starts to separate and there is a reduced rate of cooling, during which the composition of the liquid moves from b to e and the composition of the solid from d to c. At c, the liquid completely solidifies to a solid solution of that composition; thereafter, the rate of cooling increases. Components of such systems can be separated by fractional crystallisation.

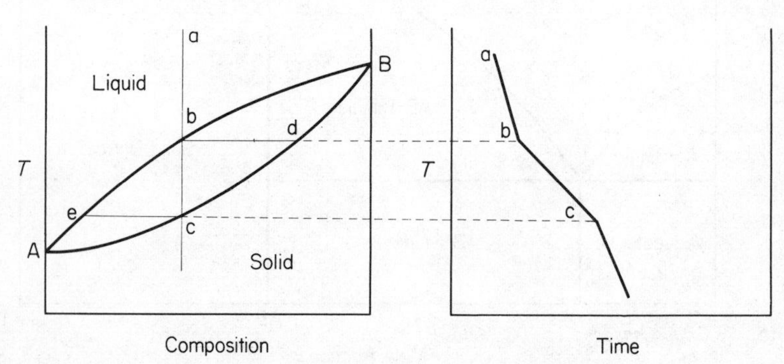

Figure T.28 Phase diagram and cooling curve for a two-component system, in which the components form a continuous series of solid solutions.

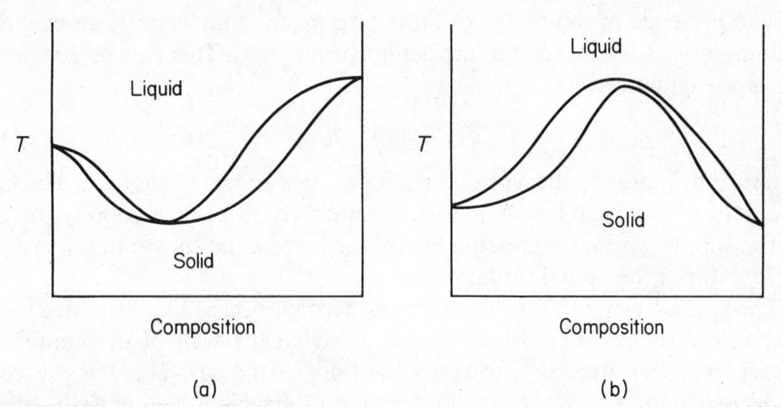

Figure T.29 Phase diagram for a two-component system, in which the components form a series of solid solutions with (a) a minimum melting point and (b) a maximum melting point.

Examples: Cu–Ni; Co–Ni; AgCl–NaCl.

(b) and (c) Those in which the solid solutions have a minimum or a maximum melting point (figure T.29). At the maximum or minimum melting point, the solid solution melts to liquid of the same composition. Only one pure component can be obtained from a given melt by fractional crystallisation.

More complicated systems, in which the components are partially miscible in the solid phase, are known; this equilibrium may be superimposed on the above equilibria.

Examples: minimum m.p., KCl–KBr and Na_2CO_3–K_2CO_3; maximum m.p., D-carvoxime–L-carvoxime.

See also Bo, F & C, F & J, M & P.

V

van der Waals' equation

van der Waals' equation is a modification of the ideal equation of state for a real gas. The failure of the ideal gas equation:

$$PV = nRT \tag{V.1}$$

to give account of the behaviour of real gases is due to (a) the volume occupied by the gas molecules and (b) the attractive forces between molecules.

van der Waals' equation

The presence of molecules of finite size means that there is an excluded volume not available for the molecules to move in. This can be accounted for in the equation

$$P(V - nb) = nRT \tag{V.2}$$

where $b/\text{m}^3\ \text{mol}^{-1}$, the volume excluded per mole of gas, is a constant characteristic of each gas. As b is determined empirically, a good correction to the simple gas law expression is obtained. It can be shown that b is equal to four times the actual molar volume.

The second van der Waals correction term concerns the attractive forces among molecules. A molecule about to strike the wall of the containing vessel has a net attraction towards the bulk of the gas. This has the effect of decreasing the impact of the molecule on the wall and so of reducing the pressure. The force exerted on such a molecule is proportional to (n/V), i.e. the number of moles of gas per unit volume; the number of molecules about to strike the wall at a given instant is also proportional to (n/V). Thus the total attractive force is proportional to $(n/V)^2$. Thus the corrected pressure is $(P + n^2a/V^2)$ and the corrected, or van der Waals, equation is

$$\left(P + \frac{n^2a}{V^2}\right)(V - nb) = nRT \tag{V.3}$$

At low pressures, V is large, a/V^2 is small and b is small compared with V, and so equation (V.3) becomes equation (V.1) and the real gas is exhibiting ideal behaviour. At slightly higher pressures, $b \ll V$ and equation (V.3) becomes

$$PV \approx nRT - n^2a/V$$

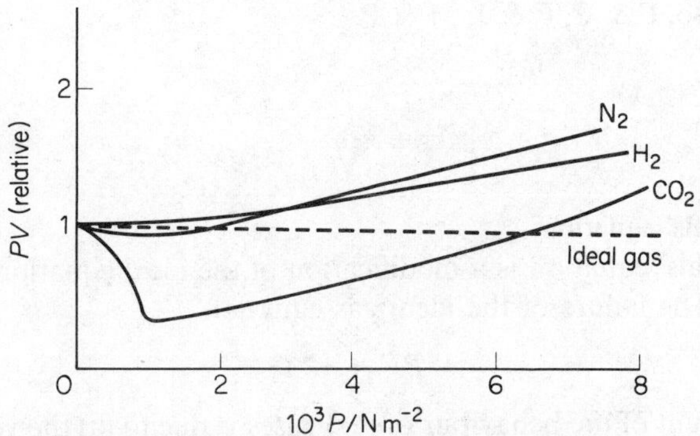

Figure V.1 Variation of PV with P for 1 mole of gas.

i.e. PV is less than RT; thus the minimum in the PV–P curve for real gases (figure V.1) is due to the cohesive forces. At very high pressures, b is now significant, but a/V^2 is small compared with P and, hence, equation (V.3) becomes

$$PV \approx RT + Pb$$

i.e. PV is now greater than RT.

The constants a, b and R are related to the critical pressure, volume and temperature of the gas:

$$b = \tfrac{1}{3}V_c;\ a = 3P_cV_c^2;\ R = \frac{8P_cV_c}{3T_c}$$

Values of a and b for some simple molecules are given in table V.1.

Table V.1. van der Waals constants

	$10^3\ a/\mathrm{N\ m^4\ mol^{-2}}$	$10^5\ b/\mathrm{m^3\ mol^{-1}}$
He	3.4	2.37
H_2	25.3	2.66
O_2	131.7	3.18
CO_2	466.1	4.27
H_2O	553.6	3.05

van't Hoff isochore

The van't Hoff isochore expresses the temperature variation of the *equilibrium constant* (q.v.) in terms of $\Delta H^\ominus$. Differentiation of the *van't Hoff isotherm* (q.v.), written in the form

$$\ln K = -(\Delta G^\ominus/RT)$$

with respect to T gives

$$\left(\frac{\partial \ln K}{\partial T}\right)_P = -\frac{1}{RT}\left(\frac{\partial(\Delta G^\ominus)}{\partial T}\right)_P - \frac{\Delta G^\ominus}{R}\left(\frac{\partial(1/T)}{\partial T}\right)_P$$

$$= \frac{\Delta S^\ominus}{RT} - \frac{\Delta G^\ominus}{RT^2} = \frac{\Delta H^\ominus}{RT^2}$$

Assuming that $\Delta H^\ominus$ is independent of temperature, the equation may be integrated:

$$\log K = -\frac{\Delta H^\ominus}{2.303\,R}\frac{1}{T} + \text{constant}$$

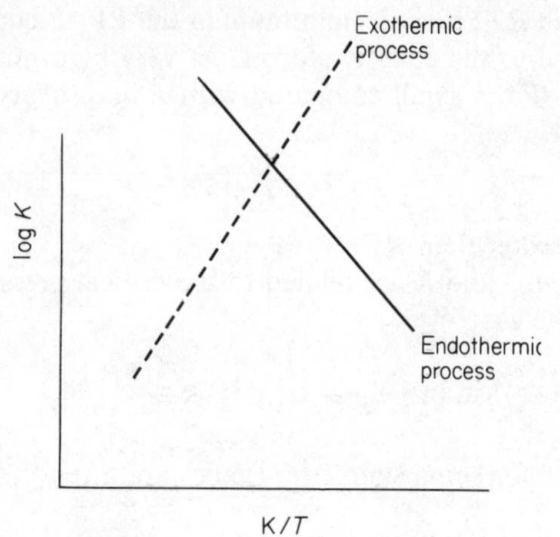

Figure V.2 Graph of log K against K/T.

Thus the slope of the straight line of log K against T^{-1} (figure V.2) is $-\Delta H^{\ominus}/2.303\ R$, from which $\Delta H^{\ominus}$ can be calculated.

Alternatively, if K_1 and K_2 are the equilibrium constants at T_1/K and T_2/K, respectively, then

$$\ln \frac{K_1}{K_2} = \frac{\Delta H^{\ominus}\ \Delta T}{R\ T_1 T_2}$$

Since $\Delta H^{\ominus}$ is a function of temperature, then the power series for *heat capacities* (q.v.) should be used, giving

$$\ln K = -\frac{\Delta H_0^{\ominus}}{RT} + \frac{\Delta a}{R} \ln T + \frac{\Delta b\ T}{2R} + \frac{\Delta c\ T^2}{6R} - \frac{J}{R}$$

The similarity of the isochore to the *Clausius–Clapeyron equation* (q.v.) is not accidental. Equations for solubility, for reaction rate constants, etc., have similar form. This is to be expected, since all these physical properties depend on the system surmounting an energy barrier (heat of reaction, vaporisation, solution, etc.) and, hence, they all ultimately depend on the *Boltzmann distribution law* (q.v.). This law is an exponential expression relating the number of molecules, energy states and temperature, and this gives rise to a logarithmic temperature dependence.

The isochore may be used to determine the standard *enthalpy* (q.v.) change from a knowledge of K at different temperatures, or to calculate

K at different temperatures from a knowledge of K at one temperature and $\Delta H^\ominus$.

van't Hoff isotherm

The van't Hoff isotherm is an equation by which the *free energy change* (q.v.) of a reaction can be calculated from a knowledge of the *equilibrium constant* (q.v.) (or $\Delta G^\ominus$) and the initial and final *activities* (q.v.) (concentrations, or partial pressures) of reactants and products. For the general reaction:

$$aA + bB \longrightarrow lL + mM$$

$$\Delta G = \sum_{\text{products}} n_i\mu_i - \sum_{\text{reactants}} n_i\mu_i$$

$$= l\mu_L + m\mu_M - a\mu_A - b\mu_B$$

Since the definition of *chemical potential* (q.v.) is

$$\mu_i = \mu_i^\ominus + RT\ln a_1$$

it follows that

$$\Delta G = l\mu_L^\ominus + m\mu_M^\ominus - a\mu_A^\ominus - b\mu_B^\ominus + RT\ln\frac{a_L^l\, a_M^m}{a_A^a\, a_B^b}$$

$$= \Delta G^\ominus + RT\ln\frac{a_L^l\, a_M^m}{a_A^a\, a_B^b}$$

where the standard free energy change is defined as

$$\Delta G^\ominus = -RT\ln K$$

and where a_A, a_B, a_L and a_M are the initial and final activities of reactants and products in non-standard states (i.e. in the chemical reaction) and K is the thermodynamic equilibrium constant for the reaction. For reactions in the gaseous phase, assuming ideal behaviour, the equation may be written

$$\Delta G = -RT\ln K_p + RT\ln\frac{P_L^l\, P_M^m}{P_A^a\, P_B^b}$$

where P_A, P_B, P_C and P_D are the initial and final partial pressures of reactants and products, respectively, in the reaction.

Consider the synthesis of ammonia at 298 K, when each of the gases is present at a partial pressure of 5 atm (i.e. $5 \times 101\,325\ \mathrm{N\,m^{-2}}$). From tabulated data $\Delta G_f^\ominus$ (NH_3, g, 298 K) $= -16.64\ \mathrm{kJ\,mol^{-1}}$ and since $\Delta G^\ominus = -RT\ln K_p$, it follows that $K_p = 82.4\ \mathrm{atm^{-1}} = 8.093\ \mathrm{m^2\,N^{-1}}$. Thus, for the reaction conditions specified,

$$\Delta G = -16.64 + 8.314 \times 10^{-3} \times 298 \times 2.303 \times \log 5/25$$
$$= -20.62 \text{ kJ mol}^{-1}$$

The change in free energy in passing from A and B completely to L and M through the equilibrium position is given by the isotherm. Such a situation can be visualised with the use of a van't Hoff equilibrium box as follows. The initial pressure of A is adjusted to that in the equilibrium mixture contained in the box, and then A is added slowly through a membrane in the side of the box, permeable only to A, to this equilibrium mixture. In a similar way, B is fed into the mixture at the correct rate to react with the slight excess of A. As L and M are formed, they are extracted through their appropriate membranes at their own equilibrium pressures and finally expanded or compressed to the final pressures of the two gases. From an evaluation of the work done in each of these processes, the same expression for ΔG may be obtained.

See also E, Wa.

Vapour pressure

Vapour pressure, p (dimensions: m l^{-1} t^{-2}; units: mmHg, N m^{-2}), or saturation vapour pressure, is the pressure at which liquid (solid) and vapour can coexist at a given temperature (from the *phase rule* q.v., $f = 1$ for a *one-component system*, q.v., in two phases at equilibrium). The v.p. is

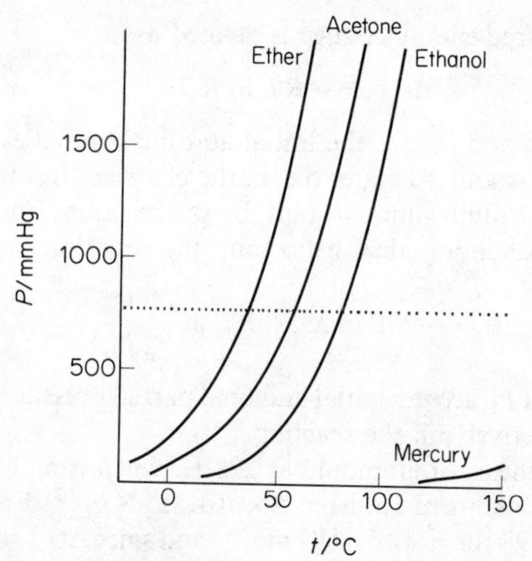

Figure V.3 Typical v.p.–temperature curves.

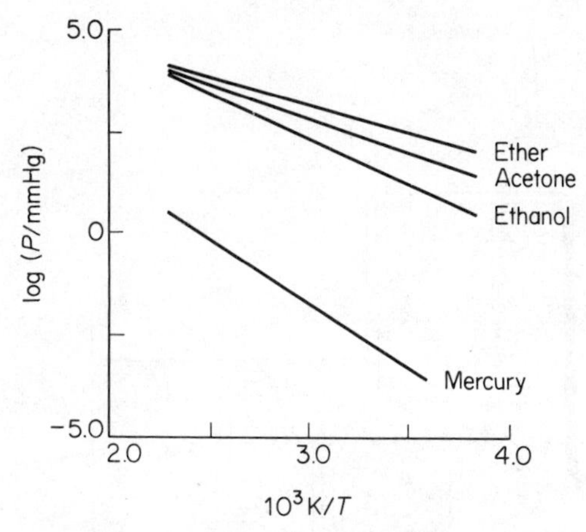

Figure V.4 log v.p.–T^{-1} plots for some liquids.

independent of the relative amounts of liquid (solid) and vapour, but depends on the temperature (figures V.3, V.4; *Clausius–Clapeyron equation*, q.v.).

The v.p. of a volatile solvent is decreased by the addition of a non-volatile solute according to *Raoult's law* (q.v.); for the addition of solute B to solvent A,

$$\frac{p_A^\ominus - p}{p_A^\ominus} = x_B = \frac{n_B}{n_A} = \frac{w_B}{M_r(B)} \frac{M_r(A)}{w_A}$$

Measurement of v.p. of pure substances

Static methods The pressure exerted when a small amount of liquid or solid is vaporised is measured at a definite temperature. The design of the manometer depends on the sensitivity required and the chemical properties of the vapour. The main types of manometer are: (1) The simple mercury manometer with one arm in contact with the vapour; this can be read to 0.01 mmHg and is suitable for pressures > 2–3 mmHg, provided the vapour does not attack mercury. (2) Magnifying manometers, e.g. two-liquid manometers for lower pressures. (3) A spoon or Bourdon gauge for low pressures, when the vapour attacks mercury. This type has a sensitivity of 0.01 mmHg. The pressure is balanced against an external pressure which can be measured with a mercury manometer.

In the simple isoteniscope (figure V.5) the v.p. of the liquid is balanced by an adjustable pressure of air so that the levels in the two arms of the U-tube are equal. The liquid must be completely degassed by pumping, and tem-

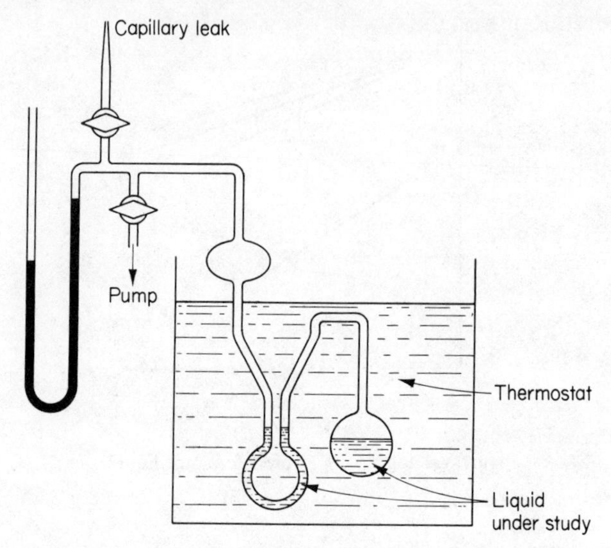

Figure V.5 Isoteniscope, for the measurement of vapour pressure.

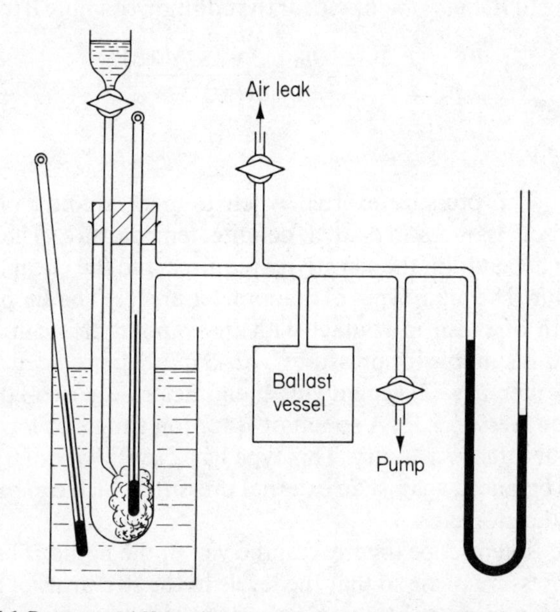

Figure V.6 Ramsay and Young's apparatus for the measurement of vapour pressure.

perature equilibrium established before the liquid levels in the two arms are adjusted. The v.p. of the liquid is given by: v.p. = barometric pressure − pressure difference on manometer.

Dynamic methods The b.p. of the liquid is determined under a fixed external pressure. In the Ramsay and Young apparatus (figure V.6), the bulb of the thermometer is at the temperature of the liquid, which is in equilibrium with vapour at the pressure in the system. The liquid on the gauze surrounding the thermometer is heated by condensation of vapour and by radiation from the outer tube, and cooled by evaporation until a steady state is reached. This is the b.p. of the liquid at a reduced pressure. The v.p. = barometric pressure − pressure difference on manometer.

Transpiration method A known volume of clean dry gas is passed over the liquid at constant temperature, and the amount carried over is measured either by chemical analysis or by freezing out with liquid air and weighing. The method is very accurate but tedious.

Effusion method This is mainly applicable to the determination of the v.p. of solids. In principle, the quantity of vapour escaping through a hole in

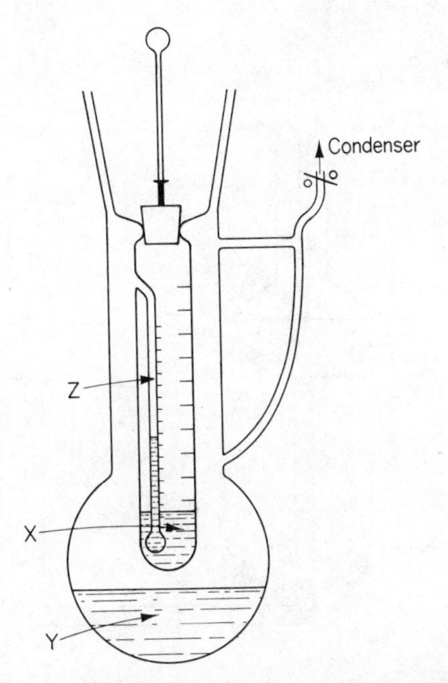

Figure V.7 Menzies apparatus for the measurement of the lowering of vapour pressure.

the thin wall of a vessel placed in a high vacuum is determined, by measuring the loss in weight of the vessel. The formula

$$N = P/(2\pi mkT)^{1/2}$$

where N is the number of molecules which hit unit area of wall per second (i.e. the number of molecules effusing through hole per second $= N \times$ the surface area of the hole) and m is the mass of the molecule, can be used to calculate the v.p. in the system. The method requires elaborate apparatus, unless the vacuum is to be broken for weighings.

Measurement of vapour pressure lowering

Static method In the Menzies method, used for the determination of molecular mass, the solution acts as its own manometer. When the solution in X and the solvent in Y (figure V.7) are in equilibrium at the b.p., the difference (l/mm) between the liquid levels in X and Z (after correction for capillarity)

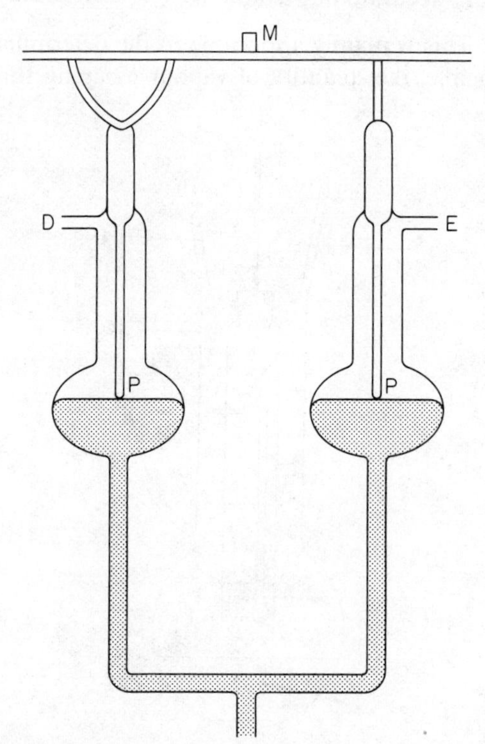

Figure V.8 Tilting manometer for the measurement of the lowering of vapour pressure.

gives $p_A^\ominus - p$ (measured in mmHg). Since, at the b.p., $p_A^\ominus = 760$ mmHg, it follows that

$$\frac{p_A^\ominus - p}{p_A^\ominus} = \frac{l\,\rho(\text{soln.})}{13.6 \times 10^3 \times 760} = \frac{w_B\, M_r(\text{A})}{w_A\, M_r(\text{B})}$$

In a second method, use is made of a more sensitive apparatus (figure V.8), in which the pressure difference is measured by the tilt of the apparatus. Initially, with arms D and E connected, the pressure is equal in both bulbs and, by means of a fine adjustment of the mercury reservoir, the glass points PP are made to coincide with their images in the mercury surfaces; this is the zero reading. The connection of D and E to the vapour above the solution and solvent, respectively, produces a slight difference in the mercury levels. The points PP are made to coincide with their images by tilting the apparatus; the extent of tilt, measured optically using a mirror M, is a measure of the pressure difference. With good temperature control, pressure differences can be measured down to 0.001 mmHg.

Dynamic methods Methods similar to those used for pure liquids have been adopted for the measurement of the v.p. of solutions. Sources of error include superheating and change of concentration due to evaporation.

Transpiration method This is essentially the same as that used for pure solvents. A typical arrangement is to pass a known volume of dry gas through a train of saturating and absorbing vessels and determine the increase in weight of the absorbers.

Isopiestic method If two vessels containing two different solutes in the same solvent are placed side by side in an evacuated and thermostatted vessel, solvent vapour will distil from the solution of higher v.p. and condense in the one of lower v.p. until equilibrium is established; both solutions now exert the same v.p. (i.e. they are isopiestic). The concentrations of the two solutions are now determined, and if the v.p. of one solution (e.g. mannitol or KCl) is known, the v.p. of the other can be calculated.

See also J & P, P (I).

Variance

See Degree of freedom.

Volume

See Partial molar volume.

W

Water system

The water system is a *one-component system* (q.v.) in which the water can exist in three phases (figure W.1). At C, liquid and vapour are indistinguishable; this is the critical point. The line AD showing the equilibrium between ice and water has a negative slope because $\rho(\text{ice}) < \rho(\text{water})$, i.e. an increase of pressure lowers the m.p. of ice.

Point M on AD is the true m.p. of ice, 0.002 30 °C at 760 mmHg pressure; note that the normal thermometric standard refers to the f.p. of water saturated with air at 760 mmHg.

The *triple point* (q.v.), A (0.0098 ± 0.001 °C and 4.58 mmHg), is now used as the fixed point on the International Temperature Scale rather than the m.p. of ice under normal atmospheric conditions.

Ice at point X, i.e. below the triple point, on heating at constant pressure (XY) passes directly into the vapour state; this process is known as sublimation. This principle is used in the modern technique of freeze-drying in the food and pharmaceutical industries. Water in the material is rapidly frozen (– 30 °C) to give ice crystals. The frozen mass is subjected to a pressure less than 4.58 mmHg and is warmed slightly to supply the heat of sublimation

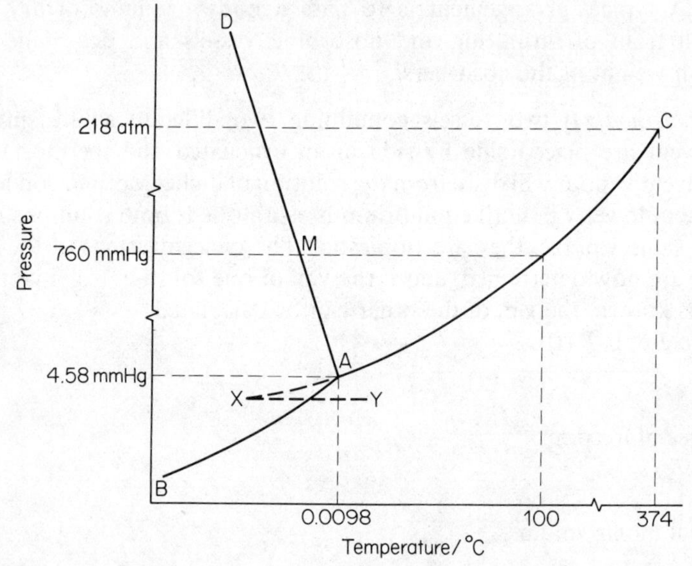

Figure W.1 Diagrammatic representation of the water system.

(temperature is still kept below -25 °C), the ice sublimes and dry material remains. This material can be reconstituted to give the original material on the addition of water. The process of freeze-drying causes no denaturation of the material, but is more costly than other methods of dehydration.

Work

Work, w or $đw$ (dimensions: ml^2t^{-2}; units: J), is defined as the product of a force and a displacement:

$$đw = F\, dl$$

Work is a form of energy transfer which may be mechanical, electrical, magnetic etc. The accepted convention is that w represents the work done by the system on its surroundings, i.e. a positive (negative) numerical value for w signifies that the system (surroundings) has done work on the surroundings (system). w is not a *state function* (q.v.); its value depends on how the process is performed (for this reason $đw$ and w are used for the work done in an infinitesimal and a finite change, respectively). The isothermal expansion of an ideal gas from state 1 (gas in half the chamber, the other half evacuated) to state 2 (gas filling the chamber completely) may be achieved in many ways, two of which (figure W.2) are: (1) opening a stopcock and allowing the gas to pass into the vacuum, no work is done on the surroundings; (2) allowing the gas to expand isothermally and reversibly against a piston until it fills the chamber; work is done by the system and, to maintain constant temperature, heat must be added to the system.

Maximum work is obtained when a process is carried out reversibly.

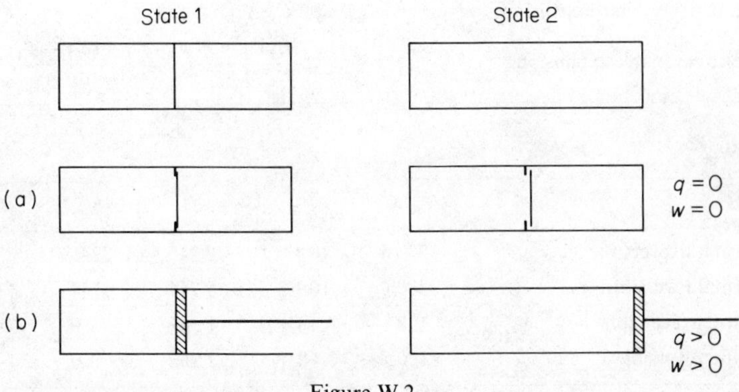

Figure W.2

Work

For an isothermal *reversible process* (q.v.), $w_{max} = -\Delta A$, the decrease in the *Helmholtz free energy* (q.v.).

Net work, w' or $đw'$, is the useful work, e.g. electrical work, that a system can provide, over and above that of expansion. The PdV work corresponds to the work 'wasted' against the confining pressure:

$$w' = w - P\Delta V = -\Delta G, \text{ or, } đw' = đw - PdV = -dG$$

The decrease in the Gibbs *free energy* (q.v.) is equal to the useful work that can be obtained from the process.

Work of expansion For the reversible expansion of n mole of an ideal gas, since $P = F/A$, it follows that

$$đw = F\,dl = PAdl = PdV$$

therefore $w = \int PdV$, i.e. the area under the P–V curve is a measure of the total work done by the gas. To calculate the maximum work, w, which for a reversible process depends only on the initial and final states, the relationship between P and V must be known, and this depends on how the process is completed.

Table W.1

Process	Work done by system
Isothermal irreversible expansion at constant pressure (P_2)	$P_2\Delta V = P_2(V_2 - V_1)$
Isothermal reversible expansion	$nRT\int_{V_1}^{V_2} dV/V = nRT \ln(V_2/V_1) = nRT \ln(P_1/P_2)$
Adiabatic irreversible expansion	$C_V\,\Delta T$
Adiabatic reversible expansion	$\dfrac{P_2V_2 - P_1V_1}{1-\gamma} = \dfrac{nR(T_2 - T_1)}{1-\gamma}$

Table W.2

Process	T_2/K	V_2	w/J	q/J	ΔU/J
Isothermal irreversible	300	10 V_1	2 245	2245	0
Isothermal reversible	300	10 V_1	5 744	5744	0
Adiabatic irreversible	192	6.4 V_1	1 347	0	− 1 347
Adiabatic reversible	120	4 V_1	2 259	0	− 2 259

Table W.1 shows the work done by the gas when the expansion is carried out reversibly or irreversibly under isothermal (see *isothermal process*) or adiabatic (see *adiabatic process*) conditions.

The values of w, ΔU and q are listed table W.2 for the various methods of expansion of 1 mole of an ideal gas, initially at volume V_1 at a pressure of 10×10^5 N m^{-2} at 300 K, to a final pressure of 10^5 N m^{-2}.

In reversible expansions, w is greater when the process is carried out isothermally than when carried out adiabatically, because ΔV is greater. For the isothermal process, w is paid for by the absorption of heat; for the adiabatic process, w is paid for by the lowering of internal energy and, hence, the temperature. For the irreversible processes, w is less than that for the comparable reversible process, since the system does not have to expand against so great an external pressure. For real gases, the above comparison between isothermal and adiabatic, reversible and non-reversible is valid, but, on account of deviations from the gas laws, the figures are not applicable.

Mechanical work The work done in lifting a weight of mass m to a height dh against the gravitational force, mg, is given by

$$đw = -mg\mathrm{d}h$$

$đw$ is negative because work is done on the system by an outside agency.

Electrical work The work obtained during the working of an electrical *cell*† of e.m.f. E at constant pressure is

$$w' = nFE$$

Work function
See Helmholtz free energy.

Z

Zeotropic mixture
A zeotropic mixture of two completely miscible liquids is a mixture which can be separated into the two pure components by *distillation* (q.v.).

See also Binary liquid mixtures.

Zeroth law of thermodynamics
Any two bodies or systems in thermal equilibrium with a third body or system are in thermal equilibrium with each other. It therefore follows that

there is some property, the temperature, whose value is the same for all systems in thermal equilibrium with each other.

The concept of temperature is based on a law of experience. Temperature is an intensive *property* (q.v.) whose value is dependent on the state of the system and is independent of the mass.

This law refers specifically to a state of equilibrium, and the implication is that the temperature can have a precise meaning only for a state of equilibrium. Experimentally, thermometers are inserted into various systems undergoing change and the recorded temperature is accepted as significant. The thermometer in such a system indicates its own temperature, which is presumed to be that of the surroundings with which it is in thermal equilibrium.

TABLES OF USEFUL DATA

Table A.I. Selected values of thermodynamic properties at 298.15 K and 1 atm pressure (101 325 N m^{-2})

The following table has been modified from various tables in the literature (e.g. Be, Dic, Ro), but stems ultimately from the comprehensive *National Bureau of Standards Circular* 500 (1952), 'Selected Values of Thermodynamic Properties', compiled by Frederick D. Rossini *et al.* Other convenient tabulations are to be found in The Chemical Rubber Company *Handbook of Chemistry and Physics*; *Lange's Handbook of Chemistry*: and *Energy Transformations in Living Matter*, by Krebs and Kornberg (Springer-Verlag, 1957).

The elements in this table are arranged alphabetically. The standard enthalpy and free energy of formation, $\Delta H_f^\ominus$ and $\Delta G_f^\ominus$, are in kJ mol^{-1}. The third law molar entropies, $S^\ominus$, and the heat capacities, $C_p^\ominus$, are expressed in J K^{-1} mol^{-1}.

c = crystalline, g = gaseous, l = pure liquid, aq = aqueous solution at unit concentration (more precisely, unit activity)

Substance	State	$\Delta H_f^\ominus$/kJ mol^{-1}	$\Delta G_f^\ominus$/kJ mol^{-1}	$S^\ominus$/J K^{-1} mol^{-1}	$C_p^\ominus$/J K^{-1} mol^{-1}
ALUMINIUM					
Al	c	0	0	28.32	24.33
Al^{3+}	aq	−525	−481	−313	
$AlCl_3$	c	−695	−637	170	89
Al_2O_3	c	−1 669.8	−1 576.4	50.99	79.0
AMMONIA and AMMONIUM					
NH_3	g	−46.2	−16.63	192.5	35.66
NH_3	aq	−80.8	−26.61	110.0	
NH_4^+	aq	−132.8	−79.5	112.8	
NH_4Cl	c	−315.4	−203.9	94.6	84
BARIUM					
Ba	c	0	0	67	26.4
Ba^{2+}	aq	−538.4	−561	13	
$BaCl_2$	c	−860.1	−811	125	75
$BaCl_2 \cdot 2H_2O$	c	−1 461.7	−1 296.2	202.9	155.2
$BaCO_3$	c	−1 218	−1 139	112	85.4
$BaSO_4$	c	−1 465	−1 353	132.2	101.8
BORON					
B	c	0	0	5.86	12.0
BF_3	g	−1 137	−1 120	254.0	50.5
BF_4^-	aq	−1 527	−1 435	167.4	
B_2H_6	g	31.4	86.6	232.0	56.4
BROMINE					
Br_2	g	30.7	3.14	245.35	36.4
Br_2	l	0	0	152	
Br^-	aq	−120.9	−102.82	80.71	−128
HBr	g	−36.2	−53·2	198.48	29.1
HBr	aq	−120.9	−102.82	80.71	

Table A.I. (*Contd.*)

Substance	State	$\Delta H_f^\ominus$/kJ mol^{-1}	$\Delta G_f^\ominus$/kJ mol^{-1}	$S^\ominus$/J K^{-1} mol^{-1}	$C_p^\ominus$/J K^{-1} mol^{-1}
CADMIUM					
Cd	c	0	0	51.5	25.89
Cd^{2+}	aq	−72.4	−77.7	−61	
$CdCl_2$	c	−389.1	−342.6	118	
CdS	c	−144.3	−140.6	71	
$CdSO_4$	c	−926.2	−820.0	137	
CALCIUM					
Ca	c	0	0	41.6	26.3
Ca^{2+}	aq	−543.0	−553.0	−55.2	
CaO	c	−635.6	−604.2	39.7	42.8
$Ca(OH)_2$	c	−986.6	−896.8	76.1	84.5
$CaCl_2$	c	−795	−750.2	113.8	72.6
$CaCO_3$	c	−1 206.7	−1 128.8	92.9	81.9
$CaSO_4$	c	−1 433	−1 320	107.0	100.0
CARBON					
C	graphite	0	0	5.694	8.64
C	diamond	1.821	2.866	2.439	6.06
CO	g	−110.52	−137.27	197.91	29.14
CO_2	g	−393.5	−394.4	213.6	37.1
CO_3^{2-}	aq	−676.3	−528.1	−53	
H_2CO_3	aq	−698.7	−623.4	191.2	
HCO_3^-	aq	−691.1	−587.1	95	
CH_4	g	−74.85	−50.79	186.2	35.71
C_2H_2	g	226.75	209.2	200.8	43.93
C_2H_4	g	52.28	68.12	219.5	43.6
C_2H_6	g	−84.67	−32.89	229.5	52.66
C_6H_6	g	82.93	129.7	269.2	81.67
C_6H_6	l	49.03	172.8	124.5	134
$H\cdot COOH$	l	−409.2	−346.0	128.9	99.0
$CH_3\cdot COOH$	l	−487	−392	160	123
CCl_4	g	−106.7	−64.2	309.4	83.5
CCl_4	l	−139	−68	214.4	131.75
$CHCl_3$	l	−132	−72	203	116
$CHCl_3$	g	−100.4	−66.9	296.5	65.8
CH_2Cl_2	g	−87.9	−58.6	270.6	51.4
CH_3Cl	g	−80.8	−57.4	234.5	40.8
CH_3OH	l	−238.7	−166.4	126.8	81.58
C_2H_5OH	l	−277.63	−174.8	160	111.5
CHLORINE					
Cl_2	g	0	0	222.95	33.9
Cl^-	aq	−167.46	−131.17	55.1	−126
HCl	g	−92.31	−95.27	186.7	29.1
HCl	aq	−167.46	−131.17	55.1	
ClO_3^-	aq	−98.3	−2.6	163	−75
ClO_4^-	aq	−131.4	−10.8	182	
CHROMIUM					
Cr	c	0	0	23.8	23.3
Cr^{3+}	aq		−215.5	−307.5	

Table A.I. (*Contd.*)

Substance	State	$\Delta H_f^\ominus$/kJ mol^{-1}	$\Delta G_f^\ominus$/kJ mol^{-1}	$S^\ominus$/J K^{-1} mol^{-1}	$C_p^\ominus$/J K^{-1} mol^{-1}
CrO_4^{2-}	aq	−894.3	−736.8	38.5	
$Cr_2O_7^{2-}$	aq	−1 460.6	−1 257.3	213.8	
COBALT					
Co	c	0	0	28.5	25.6
Co^{2+}	aq	(−67.4)	−51.0	(−155.2)	
Co^{3+}	aq		123.8		
$CoCl_2$	c	−317.1	−274.0	106.3	78.66
COPPER					
Cu	c	0	0	33.3	24.47
Cu^{2+}	aq	64.39	64.98	−98.7	
CuO	c	−155.2	−127.2	43.5	44.4
Cu_2O	c	−166.7	−146.4	100.8	70.0
$CuSO_4$	c	−769.9	−661.9	113.4	100.8
$CuSO_4 \cdot 5H_2O$	c	−2 278.0	−1 880.0	305	281.0
FLUORINE					
F_2	g	0	0	203.3	31.4
F^-	aq	−329.1	−276.5	−9.6	−123
HF	g	−268.6	−270.7	173.5	29.1
HF	aq	−329.1	−276.5	−9.6	−123
HYDROGEN					
H_2	g	0	0	130.59	28.836
H^+	aq	0	0	0	0
H_3O^+	aq	−285.84	−237.19	69.94	75.30
H_2O	g	−241.83	−228.6	188.72	33.58
H_2O	l	−285.84	−237.19	69.94	75.30
H_2O_2	l	−187.8	−113.97	(92)	
IODINE					
I_2	c	0	0	117	55.0
I^-	aq	−55.9	−51.7	109.4	−129
I_3^-	aq	−51.9	−51.5	173.6	
HI	g	25.9	1.3	206.33	29.2
HI	aq	−55.9	−51.7	109.4	−129
IRON					
Fe	c	0	0	27.2	25.2
Fe^{2+}	aq	−87.86	−84.9	−113.4	
Fe^{3+}	aq	−48.0	−10.6	−293.3	
Fe_2O_3	c	−822.2	−741.0	90.0	104.6
Fe_3O_4	c	−1 117.1	−1 014.2	146.0	
LEAD					
Pb	c	0	0	64.9	26.8
Pb^{2+}	aq	1.6	−24.3	21	
PbO (yellow)	c	−217.9	−188.5	69	48.5
$PbSO_4$	c	−918.4	−811.2	147	104
$PbCl_2$	c	−359.2	−314.0	136	77
LITHIUM					
Li	c	0	0	28.0	23.6
Li^+	aq	−278.5	−293.8	14.2	

Table A.I. (*Contd.*)

Substance	State	$\Delta H_f^\ominus$/kJ mol^{-1}	$\Delta G_f^\ominus$/kJ mol^{-1}	$S^\ominus$/J K^{-1} mol^{-1}	$C_p^\ominus$/J K^{-1} mol^{-1}
$LiCl$	c	−408.8	−383.7	55.2	
$LiBr$	c	−350.3	−339.7	69.0	
LiI	c	−271.1	−267.8		
Li_2CO_3	c	−1 215.6	−1 132.4	90.4	97.4
MAGNESIUM					
Mg	c	0	0	32.5	23.9
Mg^{2+}	aq	−462.0	−456.0	−118.0	
MgO	c	−601.8	−569.6	26.8	37.4
$MgCl_2$	c	−641.8	−592.3	89.5	71.3
$MgCO_3$	c	−1 113.0	−1 029.0	65.7	75.5
MANGANESE					
Mn	c	0	0	31.8	26.3
Mn^{2+}	aq	−219.0	−223.0	−83.7	
MnO_2	c	−519.7	−466.1	53.1	54.0
MnO_4^-	aq	−518.4	−425.1	189.95	
MERCURY					
Hg	l	0	0	77.0	27.8
Hg^{2+}	aq	174.0	164.8	−22.6	
Hg_2^{2+}	aq		153.9		
HgO (red)	c	−90.7	−58.53	72.0	45.7
$HgCl_2$	c	−230.1	−185.8	(144.3)	76.6
Hg_2Cl_2	c	−264.9	−210.66	195.8	102.0
NICKEL					
Ni	c	0	0	30.1	26.0
Ni^{2+}	aq	−64.0	−48.2	−159.0	
$NiCl_2$	c	−315.9	−272.4	107.1	77.82
NITROGEN					
N_2	g	0	0	191.5	29.12
NO	g	90.37	86.69	210.6	29.86
NO_2	g	33.85	51.84	240.4	37.9
N_2O_4	g	9.66	98.29	304.3	79.1
N_2O	g	82.05	103.59	220.0	38.7
NO_2^-	aq	−106.3	−34.5	125.1	
NO_3^-	aq	−206.6	−110.5	146	
HNO_2	aq	−118.8	−53.6		
HNO_3	aq	−206.6	−110.5	146	
HNO_3	l	−173.2	−79.9	155.6	109.9
OXYGEN					
O_2	g	0	0	205.03	29.36
O_3	g	142.7	163.4	238	38.16
OH^-	aq	−229.94	−157.3	−10.54	−133.9
PHOSPHORUS					
P	c(white)	0	0	44	23.2
P	c(red)	−18.4	−13.8	(29)	
PCl_3	g	−306.4	−286.3	311.7	
PCl_5	g	−398.9	−324.6	353	
$H_2PO_4^-$	aq	−1 302	−1 135	89	

Table A.I. (*Contd.*)

Substance	State	$\Delta H_f^\ominus$/kJ mol^{-1}	$\Delta G_f^\ominus$/kJ mol^{-1}	$S^\ominus$/J K^{-1} mol^{-1}	$C_p^\ominus$/J K^{-1} mol^{-1}
HPO_4^{2-}	aq	−1 299	−1 094	−36	
PO_4^{3-}	aq	−1 284	−1 025	−217	
H_3PO_4	aq	−1 289	1 147	176	
POTASSIUM					
K	c	0	0	64.2	29.2
K^+	aq	−251.2	−288.28	102.5	
KOH	c	−425.8	−374.5		
KF	c	−562.6	−533.1	66.6	49.1
KCl	c	−435.87	−408.32	82.7	51.5
$KClO_4$	c	−433.5	−304.2	151.0	110.2
KBr	c	−392.2	−379.2	96.4	53.6
KI	c	−327.6	−322.3	104.3	59.2
KI_3	c	−320.5	−307.5		
KNO_3	c	−492.7	−393.1	132.9	96.3
K_2CO_3	c	−1 146.1	−1 069.0		
K_2SO_4	c	−1 433.7	−1 316.4	175.7	130.1
SILICON					
Si	c	0	0	18.7	19.9
SiO_2, quartz	c	−859	−805	41.8	44.4
SiF_4	g	−1 548	−1 506	285	76
$SiCl_4$	g	−609.6	−569.9	331.4	90.8
SILVER					
Ag	c	0	0	42.70	25.49
Ag^+	aq	105.9	77.11	73.9	37.0
AgCl	c	−127.03	−109.72	96.1	50.8
AgBr	c	−99.5	−95.94	107.1	52.4
AgI	c	−62.4	−66.3	114.0	54.4
$AgNO_3$	c	−123.1	−32.2	140.9	93.1
SODIUM					
Na	c	0	0	51.2	28.4
Na^+	aq	−239.66	−261.87	60.2	
NaOH	c	−426.7	−377.0	(52.3)	80.3
NaF	c	−569.0	−541.0	58.6	46.0
NaCl	c	−411.00	−384.03	72.4	49.7
NaBr	c	−359.9	−347.7		52.3
NaI	c	−288.0	−237.2		54.4
$NaNO_3$	c	−466.7	−365.9	116.0	93.1
Na_2CO_3	c	−1 130.9	−1 048.0	136.0	110.5
Na_2SO_4	c	−1 384.5	−1 266.8	149.5	127.6
$Na_2SO_4 \cdot 10H_2O$	c	−4 324.1	−3 644.0	592.9	587.4
SULPHUR					
S	c (rhombic)	0	0	31.9	22.6
S	c (monoclinic)	0.03	0.10	32.5	23.6
S^{2-}	aq	41.8	83.6		
HS^-	aq	−17.7	12.6	61.1	
SO_2	g	−296.9	−300.4	248.5	39.7

Table A.I. (*Contd.*)

Substance	State	$\Delta H_f^\ominus$/kJ mol^{-1}	$\Delta G_f^\ominus$/kJ mol^{-1}	$S^\ominus$/J K^{-1} mol^{-1}	$C_p^\ominus$/J K^{-1} mol^{-1}
SO_3	g	−395.2	−370.4	256.2	50.6
SO_4^{2-}	aq	−907.5	−742.0	17	16
H_2SO_4	aq	−907.5	−742.0	17	16
HSO_4^-	aq	−885.8	−752.9	126.9	
SO_3^{2-}	aq	−624.3	−497.1	43.5	
H_2SO_3	aq	−608.8	−538.0	234.3	
HSO_3^-	aq	−628.0	−527.3	132.4	
H_2S	g	−20.15	−33	205.6	34
SF_6	g	−1 096.2	−991.6	290.8	
$SOCl_2$	l	−245.6	−203.4	215.7	
TIN					
Sn	c(grey)	2.51	4.6	45	25.7
Sn	c(white)	0	0	51	26.4
SnO	c	−286	−257	56	44
SnO_2	c	−587	−520	52	52.6
$SnCl_4$	l	−545	−474	259	165
ZINC					
Zn	c	0	0	41.6	25.1
Zn^{2+}	aq	−152.4	−147.2	−106.5	
ZnO	c	−348.0	−318.2	44	40.2
$ZnCl_2$	c	−415.9	−369.26	108.4	77
$ZnSO_4$	c	−978.6	−871.6	125.0	117

Table A.II. Selected average bond dissociation energies of single bonds: ΔH/kJ mol^{-1}

	H	C	O	F	Cl	Br	I
H	435	414	460	569	431	368	297
C		343	358	485	343	289	226
O			146	184	250	234	234
F				159	255	285	280
Cl					243	218	209
Br						192	180
I							151

Selected average bond dissociation energies of multiple bonds: ΔH/kJ mol^{-1}

Bond	ΔH
C=C	611
C≡C	833
C=O	741
C≡N	891

Table A.III. Selected electron affinities: $\Delta U_0/\text{kJ mol}^{-1}$ for processes $X(\text{g}) + \text{e} - \rightarrow X^-(\text{g})$

H	He	Li	Be	B	C	N	O	F
−72	54	−57	66	−15	−121	31	−142	−333
							O^-	
							844	
	Ne	Na	Mg	Al	Si	P	S	Cl
	99	−21	67	−26	−135	−60	−200	−348
							S^-	
							532	
								Br
								−324
								I
								−295

Table A.IV. Selected ionisation energies (first and second only): $\Delta U_0/\text{kJ mol}^{-1}$

	First	Second
H	1311	
He	2372	5249
Li	520	7297
Be	899.1	1758
B	800.5	2428
C	1086	2353
N	1403	2855
O	1410	3388
F	1681	3375
Ne	2080	3963
Na	495.8	4561
Mg	737.5	1450
Al	577.5	1817
Si	786.3	1577
P	1012	1903
S	999.3	2260
Cl	1255	2297
Br	1142	2080
I	1191	1842

Table A.V. Enthalpy function $(H_T^\ominus - H_0^\ominus)$

Substance	$\Delta H_f^\ominus$ /kJ mol^{-1}	$(H_T^\ominus - H_0^\ominus)$/kJ mol^{-1} at T/K					
		298.16	400	600	800	1000	1500
H_2(g)	0	8.468	11.426	17.274	23.168	29.145	44.745
O_2(g)	0	8.660	11.683	17.904	24.494	31.367	49.272
C(graphite)	0	1.053	2.103	5.013	8.710	12.864	24.325
CO(g)	−113.813	8.6719	11.6474	17.6125	23.848	30.3612	47.5252
CO_2(g)	−393.165	9.3643	13.367	22.2688	32.1724	42.7689	71.1447
H_2O(g)	−238.936	9.9064	13.3637	20.4271	27.9893	36.0159	57.9400
Methane(g)	− 66.890	10.0295	13.9034	23.2170	34.8151	48.3670	88.4079
Ethane(g)	−69.107	11.9495	17.9745	33.5389	53.3878	76.4835	144.3480
Propane(g)	−81.513	14.6942	23.2463	45.7311	74.3078	107.4033	203.5516
Ethylene(g)	60.760	10.5646	15.5268	28.1667	43.8483	61.7558	113.3864
Acetylene(g)	227.312	10.0060	14.8164	25.6354	37.6518	50.5846	85.9435
Benzene(g)	100.416	14.2298	24.1082	51.4004	86.2406	126.2020	239.9544
Toluene(g)	73.220	18.0163	30.4135	64.1575	107.1983	156.6866	298.110
o-Xylene(g)	46.426	23.3299	38.8735	79.7889	131.3190	190.5017	359.3889

Table A.VI. Free energy function $(G_T^\ominus - H_0^\ominus)/T$

Substance	$(G_T^\ominus - H_0^\ominus)/T$ /J K^{-1} mol^{-1} at T/K					
	298.16	400	600	800	1000	1500
H_2 (g)	−102.186	−110.550	−122.185	−130.482	−136.976	−148.909
O_2 (g)	−175.983	−184.565	−196.514	−205.200	−212.116	−225.133
C(graphite)	−2.16396	−3.4497	−6.1798	−8.9454	−11.594	−17.493
CO (g)	−168.824	−177.372	−189.209	−197.711	−204.430	−216.999
CO_2(g)	−182.234	−191.744	−206.012	−217.129	−226.392	−244.685
H_2O (g)	−155.276	−165.301	−178.941	−188.281	−196.723	−211.802
Methane(g)	−152.55	−162.59	−177.36	−189.16	−199.37	−221.09
Ethane (g)	−189.41	−201.84	−222.09	−239.70	−255.68	−290.62
Propane (g)	−220.62	−236.31	−263.30	−287.61	−310.03	−359.24
Ethylene (g)	−184.01	−195.02	−212.13	−226.73	−239.70	−267.52
Acetylene (g)	−167.260	−177.615	−193.774	−206.690	−217.589	−239.455
Benzene (g)	−221.46	−237.19	−266.52	−294.30	−320.37	−378.44
Toluene (g)	−259.32	−279.24	−316.06	−350.58	−382.96	−455.01
o-Xylene (g)	−274.51	−300.16	−346.48	−389.15	−428.69	−515.89

Table A.VII. Thermodynamic functions for dissolved solutes. Relation between molality, mean ionic activity and activity coefficient ($\gamma_\pm$). (m = molality, a_B = activity of solute)

Type	Example	$\gamma_\pm$	$m_\pm$	$a_B\ (=m_\pm\gamma_\pm)$
Non-electrolyte	Glucose			$m\gamma$
1:1	NaCl	$(\gamma_+\gamma_-)^{1/2}$	m	$m^2\gamma_\pm^2$
2:2	$ZnSO_4$			
3:3	$LaFe(CN)_6$			
2:1	$CaCl_2$	$(\gamma_+\gamma_-^2)^{1/3}$	$4^{1/3}m$	$4m^3\gamma_\pm^3$
1:2	K_2SO_4	$(\gamma_+^2\gamma_-)^{1/3}$	$4^{1/3}m$	$4m^3\gamma_\pm^3$
3:1	$LaCl_3$	$(\gamma_+\gamma_-^3)^{1/4}$	$27^{1/4}m$	$27m^4\gamma_\pm^4$
1:3	$K_3Fe(CN)_6$	$(\gamma_+^3\gamma_-)^{1/4}$	$27^{1/4}m$	$27m^4\gamma_\pm^4$
4:1	$Th(NO_3)_4$	$(\gamma_+\gamma_-^4)^{1/5}$	$256^{1/5}m$	$256m^5\gamma_\pm^5$
1:4	$K_4Fe(CN)_6$	$(\gamma_+^4\gamma_-)^{1/5}$	$256^{1/5}m$	$256m^5\gamma_\pm^5$
3:2	$Fe_2(SO_4)_3$	$(\gamma_+^2\gamma_-^3)^{1/5}$	$108^{1/5}m$	$108m^5\gamma_\pm^5$

Table A.VIII. Individual activity coefficients in water at 298.15 K

This table lists the values of the activity coefficients of selected individual ions at different ionic strengths (a total of 130 ions are listed by Kielland, *J. Amer. Chem. Soc.*, 59, 1675, 1937) calculated from the *Debye–Hückel activity equation* (q.v.). The parameter $\mathring{a}/10^{-10}$ m may be regarded as the effective diameter of the hydrated ion obtained by various methods

N.B. $\gamma_\pm = (\gamma_+^{\nu+}\gamma^{\nu-})^{1/\nu}$

Individual activity coefficients of ions in water

Parameter $\mathring{a}$	Ionic strength I/mol dm^{-3} 0.0005	0.001	0.0025	0.005	0.01	0.025	0.05	0.1
				Ion Charge 1				
9	0.975	0.967	0.950	0.933	0.914	0.88	0.86	0.83
6	0.975	0.965	0.948	0.929	0.907	0.87	0.835	0.80
5	0.975	0.964	0.947	0.928	0.904	0.865	0.83	0.79
4.5	0.975	0.964	0.947	0.928	0.902	0.86	0.82	0.775
4	0.975	0.964	0.947	0.927	0.901	0.855	0.815	0.77
3.5	0.975	0.964	0.946	0.926	0.900	0.855	0.81	0.760
3	0.975	0.964	0.945	0.925	0.899	0.85	0.805	0.755
2.5	0.975	0.964	0.945	0.924	0.898	0.85	0.80	0.75
				Ion Charge 2				
8	0.906	0.872	0.813	0.755	0.69	0.595	0.52	0.45
6	0.905	0.870	0.809	0.749	0.675	0.57	0.485	0.405
5	0.903	0.868	0.805	0.744	0.67	0.555	0.465	0.38
4.5	0.903	0.868	0.805	0.742	0.665	0.55	0.455	0.37
4	0.903	0.867	0.803	0.740	0.66	0.545	0.445	0.355
				Ion Charge 3				
9	0.802	0.738	0.632	0.54	0.445	0.325	0.245	0.18
5	0.796	0.728	0.616	0.51	0.405	0.27	0.18	0.115
4	0.796	0.725	0.612	0.505	0.395	0.25	0.16	0.095
				Ion Charge 4				
11	0.678	0.588	0.455	0.35	0.255	0.155	0.10	0.065
5	0.668	0.57	0.425	0.31	0.20	0.10	0.048	0.021

Table A.VIII (*Contd.*)
Values of the parameter å for selected ions

å	
	Charge 1
9	H^+
6	Li^+
5	$CHCl_2COO^-$, CCl_3COO^-
4–4.5	Na^+, IO_3^-, HCO_3^-, $H_2PO_4^-$, CH_3COO^-, $NH_2CH_2COO^-$, $C_2H_5NH_3^+$
3.5	OH^-, F^-, CNS^-, CNO^-, BrO_3^-, MnO_4^-, $HCOO^-$, H_2 citrate$^-$, $CH_3NH_3^+$
3	K^+, Cl^-, Br^-, I^-, CN^-, NO_2^-, NO_3^-
2.5	NH_4^+, Ag^+
	Charge 2
8	Mg^{2+}
6	Ca^{2+}, Cu^{2+}, Zn^{2+}, Sn^{2+}, Mn^{2+}, Fe^{2+}, Ni^{2+}, Co^{2+}
5	Sr^{2+}, Ba^{2+}, Cd^{2+}, Hg^{2+}
4.5	Pb^{2+}, CO_3^{2-}, SO_3^{2-}, $(COO^-)_2$, H citrate^{2-}
4	Hg_2^{2+}, SO_4^{2-}, $S_2O_3^{2-}$, CrO_4^{2-}, HPO_4^{2-}
	Charge 3
9	Al^{3+}, Fe^{3+}, Cr^{3+}, Ce^{3+}
5	Citrate^{3-}
4	PO_4^{3-}, $Fe(CN)_6^{3-}$, $Cr(NH_3)_6^{3+}$
	Charge 4
11	Ce^{4+}, Sn^{4+}
5	$Fe(CN)_6^{4-}$

Table A.IX. Solubility products

The data listed are the true thermodynamic solubility products at 298.15 K. They are tabulated in the form α, β defined by $K_s = \alpha\, 10^{-\beta}$. The units of K_s are determined by ν (the total number of moles of ions formed from 1 mole of salt) and are of the form $(\text{mol dm}^{-3})^{\nu}$.

Species	α	β	Species	α	β
AgBr	7.7	13	$Fe(OH)_2$	2.2	15
AgCl	1.56	10	$Fe(OH)_3$	2.5	39
AgCNS	1.2	12	Hg_2Br_2	5.2	23
Ag_2CrO_4	1.9	12	Hg_2Cl_2	1.3	18
AgI	8.3	17	$MgCO_3$	3.5	8
$BaCO_3$	5.0	9	$MnCO_3$	2.0	11
$BaSO_4$	1.0	10	$PbBr_2$	4.0	5
$CaCO_3$	4.8	9	$PbCl_2$	1.6	5
CaC_2O_4	1.3	9	PbI_2	6.7	9
$CaSO_4$	1.0	5	$PbCrO_4$	1.6	14
CdS	1.0	28	$PbSO_4$	1.6	8
$CdCO_3$	5.2	12	$SrCO_3$	1.1	10
CuBr	5.9	9	$SrSO_4$	3.2	7
CuCl	1.9	7	$ZnCO_3$	2.1	11
CuI	1.4	12	ZnS	1.6	23

Table A.X. Ionisation constant in aqueous solution at 298 K

Acids

	pK_{a_1}	pK_{a_2}	pK_{a_3}
Acetic acid	4.756		
Ammonium ion	9.245		
Benzoic acid	4.203		
n-Butyric acid	4.82		
Carbonic acid	6.352	10.329	
Chloroacetic acid	2.865		
Formic acid	3.752		
Glutamic acid	2.162	4.288	9.387
Glycine	2.350	9.780	
Maleic acid	1.921	6.225	
Malonic acid	2.847	5.696	
4-Nitrobenzoic acid	3.43		
Oxalic acid	1.271	4.266	
Phenol	9.99		
Phosphoric acid	2.148	7.198	12.32
o-Phthalic acid	2.95	5.408	
Propionic acid	4.874		
Trimethylacetic acid	5.032		

Bases

	pK_a
Ammonia	9.245
Aniline	4.61
Dimethylamine	10.73
Diphenylamine	0.79
Methylamine	10.66
Piperidine	11.12
Pyridine	5.18
Trimethylamine	9.81

Table A.XI. Recommended values of physical constants

Physical constant	*Symbol*	*Value*
acceleration due to gravity	g	$9.81\ m\ s^{-2}$
Avogadro constant	N_A	$6.022\ 52 \times 10^{23}\ mol^{-1}$
Bohr magneton	μ_B	$9.273\ 2 \times 10^{-24}\ A\ m^2\ (J\ T^{-1})$
Boltzmann constant	k	$1.380\ 54 \times 10^{-23}\ J\ K^{-1}$
charge to mass ratio	e/m	$1.758\ 796 \times 10^{11}\ C\ kg^{-1}$
Curie	Ci	37.0×10^9 disintegrations per second
electronic charge	e	$1.602\ 10 \times 10^{-19}\ C$
Faraday constant	F	$9.648\ 70 \times 10^4\ C\ mol^{-1}$
gas constant	R	$8.314\ 3\ J\ K^{-1}\ mol^{-1}$
gravitational constant	G	$66.7 \times 10^{-12}\ m^3\ kg^{-1}\ s^{-2}$
'ice-point' temperature	T_{ice}	273.150 K

Table A.XI. (*Contd.*)

Physical constant	*Symbol*	*Value*
molar volume of ideal gas at s.t.p.	V_M	$2.241\,36 \times 10^{-2}$ m^3 mol^{-1}
permeability of a vacuum	μ_0	$4\pi \times 10^{-7}$ kg m s^{-2} A^{-2} (H m^{-1})
permittivity of a vacuum	ε_0	$8.854\,185 \times 10^{-12}$ kg^{-1} m^{-3} s^4 A^2 (F m^{-1})
Planck constant	h	$6.625\,6 \times 10^{-34}$ J s
Rydberg constant	R_∞	$1.097\,373\,1 \times 10^7$ m^{-1}
standard pressure, atmosphere	P	101 325 N m^{-2}
Stefan-Boltzmann constant	σ	$5.669\,7 \times 10^{-8}$ W m^{-2} K^{-4}
triple point of water		273.16 K (exactly)
unified atomic mass constant	m_u	$1.660\,43 \times 10^{-27}$ kg
velocity of light in a vacuum	c	$2.997\,925 \times 10^8$ m s^{-1}
Wien's radiation law	$\lambda_{max} \times T$	$2.897\,8 \times 10^{-3}$ m K

Table A.XII. 20 useful equations

$\Delta U = q - w$ (p. 95)

$w = P\mathrm{d}V$ for reversible expansion of ideal gas (p. 182)

$H = U + PV$ definition of H (p. 63)

$\Delta H = \Delta U + \Delta nRT$ at constant P and T (p. 18)

$(\partial H/\partial T)_P = C_p$ (p. 118)

$[\partial(\Delta H)/\partial T]_P = \Delta C_p$ Kirchhoff equation (p. 139)

$\mathrm{d}S = \text{đ}q/T$ formal definition of S (p. 71)

$$\Delta S = \int_{T_1}^{T_2} C_p \,\mathrm{d}T/T \text{ at constant } P \quad \text{(p. 74)}$$

$\Delta S = nC_V \ln(T_2/T_1) + nR \ln(V_2/V_1)$ for an ideal gas (p. 74)

$$S = k \ln W = N_A k \ln Q + N_A kT \left(\frac{\partial \ln Q}{\partial T}\right)_P \quad \text{(p. 165)}$$

$G = H - TS$ definition of G (p. 96)

$\Delta G = \Delta H - T\Delta S$ at constant T (p. 99)

$\mu = \mu^\ominus + RT \ln a$ (p. 31)

$$\Delta G = \Delta G^\ominus + RT \ln \frac{\Pi a(\text{products})}{\Pi a(\text{reactants})} \text{ van't Hoff isotherm} \quad \text{(p. 235)}$$

$(\partial \ln K/\partial T)_P = \Delta H^\ominus/RT^2$ van't Hoff isochore (p. 233)

$$\frac{\mathrm{d}P}{\mathrm{d}T} = \frac{\Delta H}{T\Delta V} \text{ Clausius–Clapeyron equation} \quad \text{(p. 32)}$$

Table A.XII. (*Contd.*)

$$\Delta H - \Delta G = -T\left(\frac{\partial(\Delta G)}{\partial T}\right)_P \quad \text{Gibbs–Helmholtz equation} \qquad \text{(p. 116)}$$

$$\Delta S^{\ominus} = -\left(\frac{\partial(\Delta G^{\ominus})}{\partial T}\right)_P = R\left(\frac{\partial(T \ln K)}{\partial T}\right)_P \qquad \text{(p. 77)}$$

$$\mathrm{pH} = \mathrm{p}\overline{K} + \log c(\mathrm{A}^-)/c(\mathrm{HA}) \quad \text{Henderson equation} \qquad \text{(p. 123)}$$

RECOMMENDED REFERENCE BOOKS AND TEXTBOOKS

Ad — Adam, N.K. (1956). *Physical Chemistry*, Clarendon Press, Oxford

An — Andrews, F.C. (1963). *Equilibrium Statistical Mechanics.* Wiley, New York

Bar — Barrow, G.M. (1973). *Physical Chemistry*, McGraw-Hill, New York

Be — Bent, H.A. (1967). *The Second Law*, Oxford University Press, New York

Bo — Bowden, S.T. (1938). *The Phase Rule and Phase Reactions*, Macmillan, London

Br — Britton, H.T.S. (1956). *Hydrogen Ions*, Chapman and Hall, London

C — Clark, W.M. (1928). *The Determination of Hydrogen Ions*, Williams and Wilkins, Baltimore

Da — Dasent, W.E. (1970). *Inorganic Energetics*, Penguin Books, Harmondsworth

D & J — Davies, C.W. and James, A.M. (1976). *Dictionary of Electrochemistry*, Macmillan, London

Dnb — Denbigh, K.G. (1971). *The Principles of Chemical Equilibrium*, Cambridge University Press

Dic — Dickerson, R.E. (1969). *Molecular Thermodynamics*, Benjamin, New York

E, G, H & R — Eggers, D.F., Gregory, N.W., Halsey, G.D. and Rabinovitch, B.S. (1964). *Physical Chemistry*, Wiley, New York

E — Everett, D.H. (1972). *Introduction to the Study of Chemical Thermodynamics*, Longmans, Green, London

F & J — Ferguson, F.D. and Jones, T.K. (1966). *The Phase Rule*, Butterworths, London

F & C — Findlay, A. and Campbell, A.N. (1940). *The Phase Rule*, Longmans, Green, London

G — Glasstone, S. (1942). *Introduction to Electrochemistry*, Van Nostrand, New York

G & S — Gucker, F.T. and Seifert, R.L. (1967). *Physical Chemistry*, English Universities Press, London

Gug — Guggenheim, E.A. (1950). *Thermodynamics*, North-Holland, Amsterdam

H & O — Harned, H.S. and Owen, B.B. (1958). *The Physical Chemistry of Electrolytic Solutions*, Reinhold, New York

H, J & S — Hughes, M.N., James, A.M. and Silvester, N.R. (1970). *SI Units and Conversion Tables*, The Machinery Publishing Co., London

I — Ives, D.J.G. (1971). *Chemical Thermodynamics*, Macdonald, London

J & P — James, A.M. and Prichard, F.E. (1974). *Practical Physical Chemistry*, Longmans, London

K — Klotz, I.M. (1964). *Chemical Thermodynamics*, Benjamin, New York

Kn — Knox, J.H. (1971). *Molecular Thermodynamics*, Wiley International

L & R — Lewis, G.N. and Randall, M. (1961). *Thermodynamics.* 2nd edn (Pitzer, K.S. and Brewer, L.), McGraw-Hill, New York

McG — McGlashan, M.L. (1970). *Physico-Chemical Quantities and Units*, Royal Institute of Chemistry, London

M & P — Maron, S.H. and Prutton, C.F. (1965). *Principles of Physical Chemistry*, Collier-Macmillan, New York

Mo — Moore, W.J. (1972). *Physical Chemistry*, Longmans, London

N — Nash, L.K. (1970). *Elements of Statistical Thermodynamics*, Addison-Wesley, Reading, Mass.

P — Partington, J.R. (I, 1949; II, 1951; III, 1952). *An Advanced Treatise on Physical Chemistry*, Longmans, London

R — Rice, O.K. (1967). *Statistical Mechanics, Thermodynamics and Kinetics*, Freeman, San Francisco

R & S — Robinson, R.A. and Stokes, R.H. (1970). *Electrolyte Solutions*, Butterworths, London

Ro — Rossini, F.D. (1950). *Chemical Thermodynamics*, Wiley, New York

Ru — Rushbrooke, G.S. (1949). *Introduction to Statistical Mechanics*. Clarendon Press, Oxford

S & S — Strong, L.E. and Stratton, WZ. (1966). *Chemical Energy*, Chapman and Hall, London

S, G & W — Strouts, C.R.N., Gilfillan, J.H. and Wilson, A.N. (Eds.) (1962). *The Working Tools*, I and II, Oxford University Press

V — Vogel, A.I. (1961). *A Textbook of Quantitative Inorganic Chemistry*, Longmans, London

W — Wall, F.T. (1965). *Chemical Thermodynamics*, Freeman, San Francisco

War — Warn, J.R.W. (1969). *Concise Chemical Thermodynamics*, Van Nostrand Reinhold, London

Was — Waser, J. (1966). *Basic Chemical Thermodynamics*, Benjamin, New York

Wi — Wilks, J. (1961). *The Third Law of Thermodynamics*, Oxford University Press

W & W — Williams, V.R. and Williams, H.B. (1973). *Basic Physical Chemistry for Life Sciences*, Freeman, San Francisco

Wy — Wyatt, P.A.H. (1967). *Energy and Entropy in Chemistry*, Macmillan, London

SI UNITS AND CONVERSION TABLE

Physical quantity	SI unit	c.g.s. unit	Conversion factor
*Length	m	cm	10^{-2}
*Mass	kg	g	10^{-3}
*Time	s	s	1
Volume	m^3	cm^3	10^{-6}
		l	10^{-3}
Density	$kg\ m^{-3}$	$g\ cm^{-3}$	10^3
Force	†$N = kg\ m\ s^{-1} = J\ m^{-1}$	dyne	10^{-5}
Pressure	$N\ m^{-2} = kg\ m^{-1}\ s^{-2} = J\ m^{-3}$	$dyne\ cm^{-2}$	10^{-1} (1 atm = 760 mm Hg = 101 325 Nm^{-2})
Surface tension	$N\ m^{-1} = J\ m^{-2}$	$dyne\ cm^{-1}$	10^{-3}
Energy, work	†$J = kg\ m^2\ s^{-2}$	cal	4.184
		erg	10^{-7}
Molar energy, enthalpy, free energy	$J\ mol^{-1}$	$cal\ mol^{-1}$	4.184
Molar entropy, heat capacity	$J\ K^{-1}\ mol^{-1}$	$cal\ deg^{-1}\ mol^{-1}$	4.184
Electric potential	†$V = kg\ m^2\ s^{-3}\ A^{-1} = J\ A^{-1}\ s^{-1}$	$erg\ e.s.u.^{-1}$	299·8
		$erg\ e.m.u.^{-1}$	10^{-8}
*Amount of substance	mol	mol	1
Concentration	$mol\ m^{-3}$	$mol\ l^{-1}$	10^3
	$mol\ dm^{-3}$	$mol\ l^{-1}$	1
Molality	$mol\ kg^{-1}$	$mol\ kg^{-1}$	1
Mole fraction	Dimensionless quantity		
Ionic strength	$mol\ kg^{-1}$	$mol\ kg^{-1}$	1
	$mol\ m^{-3}$	$mol\ l^{-1}$	10^3
	$mol\ dm^{-3}$	$mol\ l^{-1}$	1

*Basic SI units. The others are: electric current, A; thermodynamic temperature, K; and luminous intensity, cd.
†Special names for derived units which are coherant with the SI units.